Reliability of Microtechnology

Johan Liu • Olli Salmela • Jussi Särkkä
James E. Morris • Per-Erik Tegehall
Cristina Andersson

Reliability of Microtechnology

Interconnects, Devices and Systems

Springer

Johan Liu
SMIT Center and Bionano Systems Laboratory
Department of Microtechnology and Nanoscience
Chalmers University of Technology
Kemivägen 9, SE-412 96 Göteborg
Sweden
and
Key Laboratory of New Displays and System Integration
SMIT Center and School of Mechatronics and Mechanical Engineering
Box 282, Room 316,
Mechatronics Building, Shanghai University
No 149, Yan Chang Road
Shanghai 200072, China
jliu@chalmers.se

Olli Salmela
Nokia Siemens Networks
Linnoitustie 6, FI-02600 Espoo Finland
olli.salmela@nsn.com

Jussi Särkkä
Nokia Siemens Networks
Kaapelitie 4, FI-90620 Oulu
Finland
jussi.sarkka@nsn.com

James E. Morris
Department of Electrical & Computer Engineering
Portland State University
P.O. Box 751, Portland
OR 97207-0751
USA
jmorris@cecs.pdx.edu

Per-Erik Tegehall
Swerea IVF
Box 104, SE-431 22 Mölndal
Sweden
per-erik.tegehall@swerea.se

Cristina Andersson
Department of Microtechnology and Nanoscience
Chalmers University of Technology
Kemivägen 9, SE-412 96 Göteborg
Sweden
cristina.andersson@chalmers.se

ISBN 978-1-4419-5759-7 e-ISBN 978-1-4419-5760-3
DOI 10.1007/ 978-1-4419-5760-3
Springer New York Dordrecht Heidelberg London

Library of Congress Control Number: 2011920685

Printed on acid-free paper

Springer is part of Springer Science+Business Media (www.springer.com)

Foreword

As a society, our dependency on electronics products permeates every corner of the world, at an ever accelerated pace.

As an industry, we have been and will continue to place the highest priority on product reliability, while facing increasingly demanding customers and mounting competitive pressures. It is widely recognized that product reliability issues can result in inconvenience in some cases and catastrophe in others.

It is, therefore, a matter of the utmost importance that we educate our college students and practicing engineers on the reliability of microtechnology – its theory and practical applications.

The issue of reliability, however, is complicated by the wide variety of application environments and requirements, which give rise to different stress conditions (thermomechanical, dynamical, electrochemical, electrical, etc.). The picture is further complicated by the constant emergence of new applications (therefore the associated use and environmental conditions), new product designs, new materials, and new processes.

We are fortunate to have a few dedicated experts who can lead and guide us through the critical but complex issues associated with electronics reliability. We have learned a great deal from them at conferences and workshops; however, a comprehensive text book has long been awaited. This book is very timely indeed.

September, 2010

Dongkai Shangguan, Ph.D., MBA
Vice President – Flextronics
Visiting Professor – HUST, China
Fellow – IEEE

Preface

This book serves as a teaching material concerning reliability of microtechnology and covers topics from devices to systems in the final year of undergraduate and first year of graduate education including questions and answers for self-study. The book is also useful for reliability engineers for reliability assessment, modeling, and quality control purposes. The book includes reliability issues of interconnects, component up to system level. The methodology of reliability concept is addressed in the first chapters and followed by general failure mechanisms including specific failure modes in solder and conductive adhesives. Accelerated testing, interconnect, component, and system-level reliability are described also in detail as well as the reliability design for manufacturability. Finally, quality and reliability management issues as well as characterization tools for reliability are described.

Gothenburg
September, 2010

Johan Liu
Member of the Royal Swedish Academy
of Engineering Sciences (IVA)
Fellow IEEE
Professor Chalmers University of Technology, Sweden
Special Recruited Professor Shanghai University, China

Contents

Chapter 1
Introduction to Reliability and Its Importance

1.1 Introduction

Reliability engineering is becoming a multidisciplinary science. In earlier days, reliability engineering was considered as equal to applied probability theory and statistics. Nowadays, the reliability research area has been clearly subdivided into smaller entities. The research topics may be divided by the methodology applied; mathematics-based approaches have a long history, especially in reliability analysis of large systems, while physics-based approaches are being introduced, especially in component level studies. New concepts in mathematics are swiftly being introduced to reliability engineering. These include, for example, fuzzy logic [1] and Petri Nets [2]. Physical reliability science has benefited from the increasing computing power that has enabled accurate modeling of complex structures [3–5].

The specialization trend has many desired implications: The accuracy of reliability predictions is getting better [6], and therefore the required safety margins have become smaller. Research in specialized areas also has a tendency to create better results than those achieved when working on a wide research area. One might even state that through specialization reliability is becoming a science instead of being more or less a philosophy.

However, specialization has also some negative impacts. The most obvious one is that as reliability specialists are nowadays focusing on their area of interest only, the interaction between different research topics is getting weaker. In a worst-case scenario, reliability experts cannot understand anymore the neighboring research area problems. Now, it is already evident that component level reliability analysis cannot be fully applied to higher system hierarchy-level reliability considerations. On the other hand, the component-level reliability requirement should originate from system-level requirements.

The present compendium presents a holistic approach of the reliability issues in interconnects, devices, and up to systems in microtechnology. It basically discusses the fundamentals in this field and applications specifically to electronics and MEMS fields.

J. Liu et al., *Reliability of Microtechnology: Interconnects, Devices and Systems*,
DOI 10.1007/978-1-4419-5760-3_1, © Springer Science+Business Media, LLC 2011

References

1. C.E. Pelaez and J.B. Bowles, "Applying Fuzzy Cognitive Maps Knowledge Representation to Failure Modes Effects Analysis", Proc. of the IEEE Annual Symposium on Reliability and Maintainability, 1995, 450–456.
2. M. Silva and S. Velilla, "Error Detection and Correction on Petri net models of Discrete Events Control Systems", Proc. IEEE Int. Symposium Circuits and Systems, 1985, 921–924.
3. IEEE Standard Methodology for Reliability Prediction and Assessment for Electronic Systems and Equipment, #1413–1998, IEEE, 1998.
4. IEEE Guide for Selecting and Using Reliability Predictions Based on IEEE 1413, #1413.1–2002, IEEE, 2002.
5. R. Darveaux, "Effect of Simulation Methodology on Solder Joint Crack Growth Correlation", Proc. of the 50th Electronics Components and Technology Conference, Las Vegas, 2000, 1048–1058.
6. J. Galloway, L. Li, R. Dunne, and H. Tsubaki, "Analysis of Acceleration Factors Used to Predict BGA Solder Joint Field Life", Proc. SMTA International, Chicago, 2001, 357–363.

Chapter 2
Reliability Metrology

Abstract In this chapter, first, reliability is defined. Then, different ways of modeling reliability are discussed. Empirical models are based on field data and are easy to use. Physical models address a certain failure mechanism and are used to predict wearout. Physical models may be either analytical or they may be run by computer simulations. Other useful information on reliability may be obtained by testing either test vehicles or entire products. Comparing the test results with the test results obtained, when testing similar items with field data, gives a quite good idea on which kind of field reliability performance should be anticipated. Interconnection reliability must also be taken into account when checking the reliability of a component. Many times, the actual component may not represent a large risk, whereas solder interconnection may create risks that need to be mitigated. In the end of this chapter, some statistical distributions are discussed. Especially, practical advice on how to use Weibull distribution is revealed.

2.1 The Definition of Reliability

Reliability may be defined in several ways. The definition to be used here is the commonly used definition adapted from [1]:

> Reliability is the probability that an item operating under stated conditions will survive for a stated period of time.

The above definition has its roots in military handbook MIL-STD-721C [2] and is valid for nonrepairable hardware items. The "item" may be a component, a subsystem, or a system. If the item is software instead of hardware, the definition will be somewhat different [3].

2.2 Empirical Models

Component-level reliability analysis conventions have their background in the military and space industries. As the components used in these applications were clearly safety critical, it was necessary to create qualification criteria and reliability

J. Liu et al., *Reliability of Microtechnology: Interconnects, Devices and Systems*,
DOI 10.1007/978-1-4419-5760-3_2,

prediction methods [4]. These reliability prediction models were typically based on large field failure databases. The empirical models give a generic estimate for a certain component or technology. Although also being based on empirical data, the effect of field environment was taken into account by "factors" responsible for the degradation effects related to temperature, voltage, or some other stress factor. The temperature dependence was taken into account by the so-called Arrhenius equation [5] that was originally developed when modeling the rate of chemical reactions.

However, although since the early 1970s the failure rates for micro devices have fallen approximately 50% every 3 years [6] and the handbook models were updated on the average every 6 years, the models became overly pessimistic. Finally, in 1994, the US Military Specifications and Standards Reform initiative led to the cancellation of many military specifications and standards [7]. This, coupled with the fact that the Air Force had redirected the mission of the Air Force Research Laboratory (the preparing activity for MILHDBK-217) away from reliability, resulted in MIL-HDBK-217 becoming obsolete, with no government plans to update it.

The cancellation of MIL-HDBK-217 was by no means the end of empirical models. Several similar kind of handbooks still exist, such as Bellcore Reliability Prediction Procedure [8], Nippon Telegraph and Telephone (NTT) procedure [9], British Telecom Handbook [10], CNET procedure [11], and Siemens procedure [12]. The predicted failure rates originating from different standards may, however, deviate from each other [13]. Empirical models can, in principle, also take into account early failures and random failures, which is not usually the case when considering physical models. Empirical models are also easy to use.

2.3 Physical Models

Each physical model [14, 15] is created to explain a specific failure mechanism. First, the testing is performed, the failed samples are analyzed, and the root cause for the failures is discovered [16]. Then, a suitable theory that would explain the specific failure mechanism is selected and used to calculate the acceleration factor and the predicted meantime-to-failure (MTTF) value. This means, that the acceleration factor relevant to the failure mechanism is not usually known prior to the testing and analysis of the root cause.

Physical modeling may be based either on an analytical model or on finite element analysis (FEA) simulations. Physical models are most widely applied in solder joint fatigue modeling. Some other phenomena that have been studied by physical models are electromigration [17], and other thermally induced failure mechanisms [18]. When applying physical models, it is possible to study the effects of material properties, dimensions, and field environment. The problem lies in the large parameter sensitivity of these models. Many models are applying exponential or power equations. The generic solutions to second-order differential equations, usually solved by running FEA simulations, are of exponential type. Therefore, even slightly inaccurate parameter values may result in tremendous errors. Despite this fact, proper error estimates are given far too seldom, although some examples of this do exist [19, 20].

Another aspect that may possibly degrade the level of confidence toward predictions based on physical models is the fact that the models are developed in a well-controlled laboratory environment and there is little reliability data originated in a real field environment [21]. Presently, there are still situations in which no model that would explain the failure mechanism encountered can be found. In those cases, no prediction based on physical models can be given. Physical models usually address to wearout phenomena and, therefore, are of little value if early failures or random failures are in question. The exception to this is overstress events that can be analyzed by stress–strength analysis. Also, methods to assess early failures of defective subpopulations are being developed [22].

2.4 Reliability Information

As discussed earlier in this chapter, there are different ways to estimate the reliability of microsystem components. In order to be able to evaluate the usefulness of such estimates, there should be some key criteria selected for this. One key issue is how much we can rely on the reliability data. Reliability prediction with no correlation to the actual field performance is of little value. It is also vital that the data are available at times when it is useful. After its service life, it is possible, at least in theory, to know exactly the reliability performance of a certain component population. However, this information may not be very useful, as the components have already failed and there is no means anymore to affect the retrofit costs. Therefore, timely information that is based on the best knowledge available would be most desirable for the majority of engineering purposes.

In Fig. 2.1, some reliability information sources are judged based on the two aforementioned criteria: the level of confidence on the reliability information and the time span when the information is available. The graph may be somewhat subjective but should still be quite illustrative. The ranking of the methods based on the level of confidence may be open for discussion. The term "level of confidence" is used here loosely to describe how accurate or trustworthy the information is. Level of confidence should not be confused with the confidence limits or confidence intervals that have exact definitions in statistics.

When a component is selected for use in a design, the first indication of its reliability can be based on similar item data. If a similar component has already been used for several years, it is probable that in-house field failure databases can estimate the forthcoming reliability of the introduced component. If the component has not been utilized in a similar product, it is still possible to obtain some generic estimate of its reliability based on the handbooks discussed earlier. However, it should be noted that such information might be based on out-of-date data.

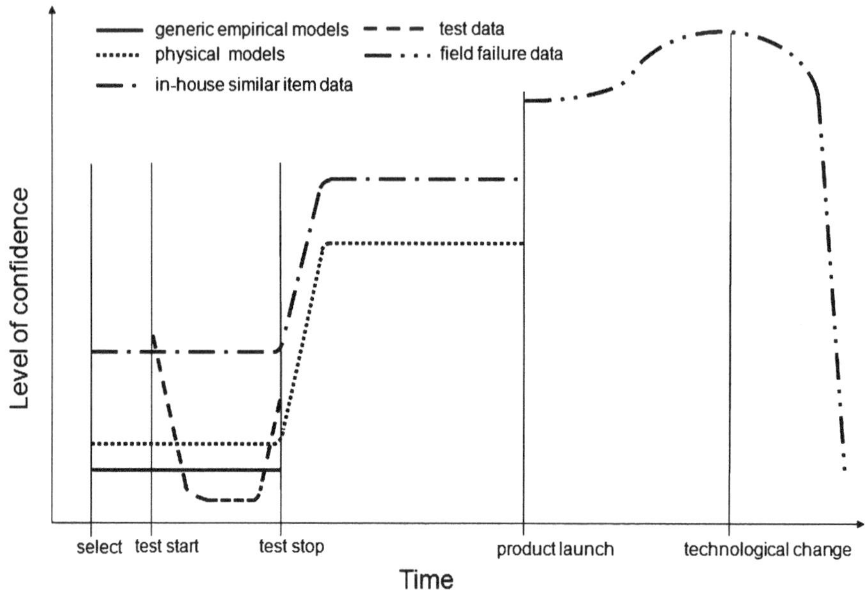

Fig. 2.1 Reliability information sources: the level of confidence of the information and the time when the information is available

If there is no field data available, physical modeling may also give an initial estimate. Physical modeling is comprised of the utilization of a suitable analytical model and/or a computer simulation analysis. As physical modeling without calibration information may not be very accurate, it is expected that in-house field data in the initial phase would be superior to physical predictions in terms of level of confidence. However, if the generic handbook values are based on old data, the physical models may give a more accurate lifetime prediction.

Only after the reliability testing has been performed is it possible to improve the quality of reliability predictions. The physical models can utilize the test results as an input (calibration data). After this, a more accurate lifetime prediction for the component can be obtained. Moreover, after the test has been concluded, it is possible to compare the test results of the component to similar items that have been tested in the same way and whose field failure data are available. This enables the reliability prediction to be based on concrete data; if the component has performed in the same way as the reference item, it is also probable that the component studied will have approximately similar field reliability behavior. If the component has performed worse than the item on which field data exist, it is expected that the field reliability performance will be somewhat worse than the reference, and vice versa.

The test itself also gives valuable information. If some early failures occur in a test it is a clear signal that there will most likely be early failures in the field as well. This is very important information, which may be very difficult to obtain unless one actually tests the component. Physical models usually predict the wearout of the components only and, therefore, may be of little value when it comes to predicting early failures. The exception to this is overstress events that can be analyzed by stress–strength analysis. Empirical models may be better at taking early failures into account. However, currently, they are not updated very often and, therefore, may be either pessimistic or they may not contain information on the new component type at all.

The shape of the test data curve resembles the bathtub curve used commonly within the reliability community. The early-failure, random-failure, and wearout regions are easily recognizable. However, as the level of confidence – instead of hazard rate – is the parameter monitored, it is expected that the shape of the curve deviates somewhat from the conventional bathtub curve. The occurrence of early failures in a test environment is a relatively reliable indicator that real concerns in the field environment are likely to take place. As the test continues, and failures occur, it may be more difficult to predict if these failures are going to be induced also by the real environment during the life span of the component. The random-failure region obtains a relatively small level of confidence value as it is expected that only a minor share of component population is going to fail during this period of time. After wearout phenomena start to occur, the confidence level is expected to rise again. This time, the level of confidence is, however, less than in case of early failures, as more time has elapsed since the test started. Therefore, it is more difficult to estimate if failures due to wearout are going to be recorded during the life span of the component.

Despite the lack of information on early failures, many times they are responsible for the majority of field failures. This is especially true when it comes to consumer products, whose expected lifetime of use is limited, and therefore, wearing out of components is not very probable. Early failures are due to design bugs, manufacturing faults, and quality problems. After the early-failure period in testing, there is usually a period of time, during which not many failures occur. This is often called the "random failure" section of the bathtub curve. As only a minor share of the equipment fail during this period, the level of confidence is usually low due to the limited number of failure observations. In order to gain an acceptable level of confidence, thousands of items should be tested [23]. This is, however, in conflict with the number of test items usually available and the limited test resources.

As discussed earlier, the information on the wearout period during the test can be used as input data for other prediction methods. After the product has been launched in the field, field failure data starts to accumulate. Ideally, field data would be the most accurate source of reliability information. Unfortunately, the field data may not always be very useful for reliability engineers. There are several reasons for this. The failure analysis is not always thoroughly performed. This is due to the fact that the primary interest of the repair personnel is to repair the product, not to analyze the cause of failure. The field data also contains some failures that are not actually due to the inherent reliability level of the components.

These failures include, for example, misuse of the product. Of course even this kind of information may be valuable, if it is considered that improving the durability of the product is needed. Also, the load history of the failed component is usually lacking, which makes it difficult to understand how the failure was actually initiated. Despite these words of caution, much can be gained if field failure data are utilized effectively. If constantly monitored, the field failure data can provide useful information on subjects of improvement. Improvements based on field data can usually be implemented during the lifetime of the product. However, field data are valid only for a limited time. Technological advances are mostly responsible for this. It may be that the reliability performance of the component improves very much when the technology gets more mature. This has occurred in conjunction with integrated circuit technology, where constant improvements take place. According to MILHDBK-217 version A, a 64-kB RAM would fail in 13 s [24].

This very pessimistic prediction is a most unfavorable example of empirical models. Nowadays, the RAM capacity is several thousand times larger than in the example given, and still, RAMs are not considered as reliability concerns. Another cause for field failure data becoming obsolete is the fact that components and component technologies have a natural life span. Due to the technology-obsolescence cycle, technologies will be replaced by some other technologies, and therefore, reliability estimates using the old technologies are of no interest.

Figure 2.2 shows a new interpretation of stress–strength distribution of a product population. The overlapping part describes the probability for the product failures and the strength of the product will gradually degrade (aging).

Figure 2.3 shows a typical approach to product total failure estimates in time, with constant component hazard rate. The failures are based on randomly failed individual products that could be a result of quality variation within the product population.

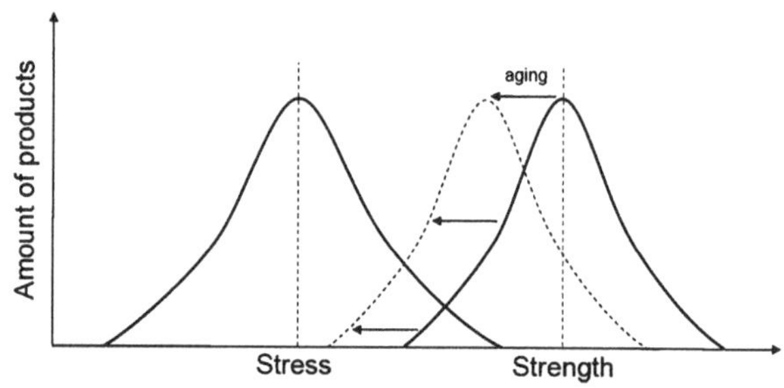

Fig. 2.2 New interpretation of stress–strength distribution of a product population

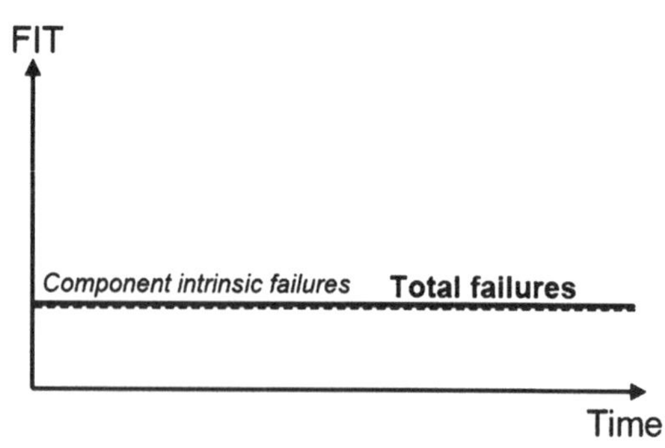

Fig. 2.3 Product failures-in-time (FIT) caused by components with constant hazard rate

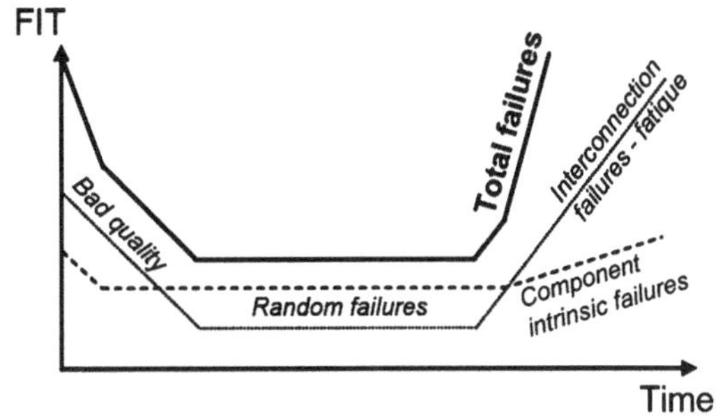

Fig. 2.4 Product failures-in-time (FIT) caused by component intrinsic and interconnection failures. Material and system aging mechanisms are involved

It gives somewhat adequate estimates of product failures, especially when the product component set is based on mature technology, the usage environment will be quite the same, and the stress levels do not exceed the strength of the product. In order to make reasonable credible predictions, the new system must be similar to well known existing system without involving significant technological risks.

However, if any of the components hazard rates starts to increase during the design lifetime of the product, the constant hazard rate approach is not relevant anymore. In practice, the total hazard rate is a sum of a time-independent constant hazard rate and a time-dependent increasing hazard rate with infant mortality region included. So the product hazard rate would look more like the so-called bathtub curve shown in Fig. 2.4. The component intrinsic failures will increase in time, which will increase the hazard rate. If the total product hazard rate is presented, the other failure modes, e.g., the interconnection failures, should be included. This addition increases the hazard rate even further.

2.5 Interconnection Reliability

In this section, interconnection reliability theories are introduced. The general reliability theories are utilized for this. Figure 2.5 shows different approaches to component reliability, where the component intrinsic and its second-level elements are shown. Approach 1 is the most often used approach, where only the component intrinsic failures are taken into account. These failures include the failures at an IC level as well as a component packaging level.

Approach 1 does not cover the failures caused by second level interconnections, namely, the solder joints or mechanically pressed joints. This approach is relevant for components with robust interconnections. In such systems, the reliability of a component, R, is written simply as:

$$R = R_{\text{component}}, \tag{2.1}$$

where $R_{\text{component}}$ is an intrinsic reliability of the component.

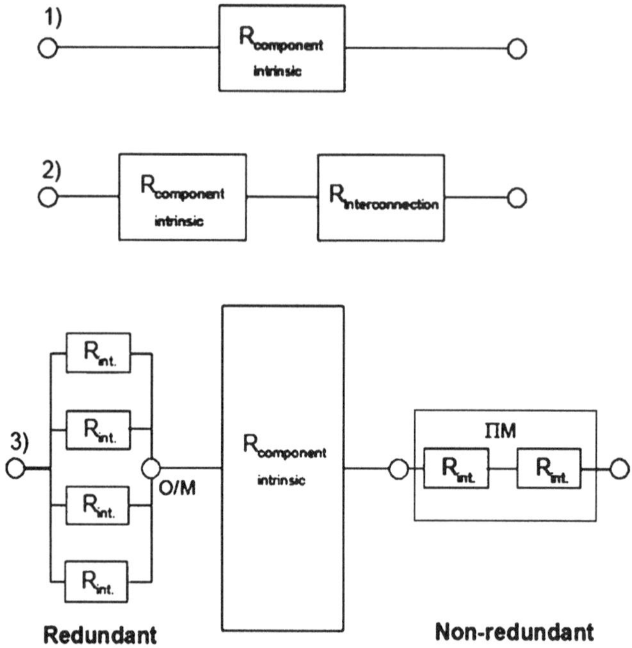

Fig. 2.5 A diagram representation of different approaches (1–3) of component reliability, divided to component intrinsic ($R_{\text{component}}$ intrinsic) and its second-level interconnection elements (R_{int}). O and M denote to the amount of elements to work to maintain the functionality of the system

Approach 2 takes into account the reliability of the second level interconnections. This approach is applicable for components whose second level interconnections are one of the main failure sources. The reliability equation for a component is then written:

$$R = R_{\text{component}} \times R_{\text{intconnection}}, \tag{2.2}$$

where $R_{\text{interconnection}}$ is the reliability of the second level interconnections of the component.

Approach 3 is based on Approach 2, but a redundancy of the interconnection elements is shown in detail. This is usually the case with many current component technologies, e.g., processors, where the supply voltage and ground are divided into many subnets. This is done to ensure the thermal and the electrical stability of a component. In spite of that, the most important interconnections, the signal inputs and outputs, do not usually have redundancy. The reliability function of a non-redundant interconnection element of Approach 3 is written as:

$$R_{\text{interconnection}} = \prod_{i=1}^{M} (R_{i,\text{interconnection}}), \tag{2.3}$$

where M is the number of the interconnections, $R_{i,\text{interconnection}}$ is a reliability of particular nonredundant interconnection element. The redundant part of Approach 3, where the interconnection elements are in parallel, can mathematically be described as:

$$R_{\text{interconnection}} = 1 - \prod_{i=1}^{M} \left(1 - R_{i,\text{redundant\_interconnection}}\right), \tag{2.4}$$

where M is the amount of parallel interconnection and $R_{i,\text{redundant\_interconnection}}$ is the reliability of each redundant interconnection element. In the redundant solder joint system, there are multiple joints that have the same function in the system to improve the long-term stability and reliability of the system. In such systems, there is M pieces of interconnections of which a fraction of interconnections, denoted as O, must work to maintain the function of the system. This is referred to as an M-modular redundant system, which can mathematically be described as:

$$R_{O/M} = 1 - \sum_{i=0}^{O-1} \left\{ \frac{M!}{(M-i)!i!} \right\} R^{i}_{\text{interconnection}} \left(1 - R_{\text{interconnection}}\right)^{M-i}, \tag{2.5}$$

where $R_{\text{interconnection}}$ is the reliability of each interconnection element. In practice, every solder joint has its individual failure expectation, which would make the reliability calculation very complex in the M-module redundant systems. In spite of that, (2.5) gives more realistic reliability expectations for redundant systems than the conservative reliability equations for interconnection elements. Figure 2.6 shows the reliability of a redundant system as a function of reliability of each interconnection element, by using (2.5). There are ten solder joints in parallel.

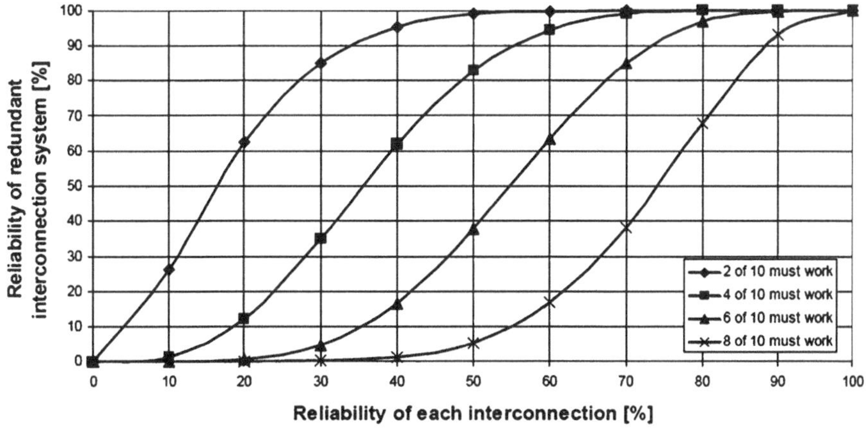

Fig. 2.6 Reliability of a redundant interconnection system as a function of the reliability of the interconnection elements. The interconnection system has ten interconnection elements in parallel, of which a fraction must work to maintain the function of the whole system. 2/10, 4/10, 6/10, and 8/10 fractions are shown

To achieve 90% reliability in the solder joint system, when two out of the ten solder joints must work to maintain the function of whole system, the reliability of one solder joint should be at least 34%. With the eight out of ten scenarios, the reliability of each solder joint should be at least 88%. As a reference, when ten solder joints are in series with respect to the reliability, a 98.95% reliability of each solder joint would be needed for 90% overall reliability.

2.6 The Levels of Interconnections

There are numerous failure mechanisms involved in electronic systems, which can be roughly divided into hardware- and software-related failures and failure mechanisms. The hardware failure mechanisms are numerous, where one of the most common hardware failure mechanisms is related to component failure. Furthermore, the component failures could be divided into component intrinsic and component interconnection (extrinsic) failures.

Microsystems are made of different levels of components or subassemblies and their connections to each other. The amount of connections reduces when getting closer to the system level, as can be seen from Table 2.1. A closer look would result in about 10^{10} subelements and interconnections within a typical digital electronic system. In practice, it would be impossible to accurately predict the reliability of such systems in every packaging and interconnection hierarchy level. Despite that, the total reliability concept should consist of every reliability element of the system. Here the emphasis is on the second level interconnections, and in more detail in the solder joints.

The typical solder joint count in a digital electronics board is between 10^4 and 10^5. Failure in any of the solder joints will degrade the functionality of the product or in the worst case stop its function entirely. So there are up to 10^5 failing elements in the digital product, which are usually left out of the risk analyses. In the most straightforward designs, where no redundancy is used, the failure in any of the components or their connections will lead instantly to system failure. In current electronic devices, this vulnerability has been taken into account, and enough redundancy has been applied to maintain the product function at least for the designed lifetime.

Table 2.1 Packaging and interconnection hierarchy and examples

Level	Description*	Example	Amount
0	Gate to gate	Transistors in ASIC chip	2×10^{9}[a]
1st	Chip to substrate	ASIC component	4,400[b]
2nd	PWB level connections	Digital electronics	$10^4 < 10^5$[a]
3rd	Connections between PWBs	Digital terminals	3,200[b]
4th	Connections between subassemblies (rack tray)	Digital electronics	$10^2 < 10^3$[b]
5th	Connections between physically separate systems	Digital electronics	$<10^2$[a]

* Brown 1998
[a] iNEMI 2004
[b] Estimated by Särkkä

Interconnection reliability analysis has not been a solid part of the product reliability concept. This is mainly due to the interconnection technologies and the set of components used in the through-hole and early surface mount technology era. Mainly in the 1970s to 1990s, the interconnection failures were not the main cause of the product failures. This can be verified from the MIL-HDBK-217F (1991), where the empirical data-based solder joint base hazard rates are given by 41 failures per 10^{12} h for through-hole assemblies and 69 failures per 10^{12} h for reflow solder joints. To put this in perspective, the hazard rates are roughly a thousand times lower than the typical component intrinsic failures for current processors. This is due to the stress-relieving leads together with relative high volume of solder and wide contact areas between lead and solder.

With the emphasis on miniaturization of the products, the components and their interconnections are getting smaller. This will lead inevitably to a situation where the interconnection failures will be playing a more significant role in product failures. New technologies are evolving continuously, and the product design time is getting shorter. Furthermore, new technologies that do not have field experience are taken in to the new product designs with accelerating pace. New technology implementation will usually mean an advantage in respect of the competitors and value adding to the customers. In order to gain the trust to the new technology, powerful and user-friendly applications for failure estimates must be taken into use. Ignoring the importance of the second-level interconnections would potentially result in catastrophic field performance at least in the very harsh usage environment and with so-called long-life high-reliability microsystems.

The standard surface mount components, which are being manufactured under standardized and mature process steps, have been in use for the past two decades. Despite the long experience of these components, their solder joint reliability performance has not been monitored under any specified or standardized procedure. With the ever increasing overall requirements for the electronics, the standard components may also become very risky. This development increases the need for a total reliability concept, where the interconnection elements are taken into account. This is emphasized with the higher risk component packages.

Before one can interpret reliability data in the literature, or compare results from different sources, it is necessary to develop a formal framework of reliability theory and definitions.

2.7 Reliability Function

Let us denote $F(t)$ to be the failure function. Then the reliability function is:

$$R(t) = 1 - F(t). \tag{2.6}$$

Let us look at a simple test: we have 1,000 flip-chip joints. After 10 h, 103 joints are broken. Then the failure function is $F(t) = 103/1{,}000 = 0.103 = 10.3\%$ at 10 h.

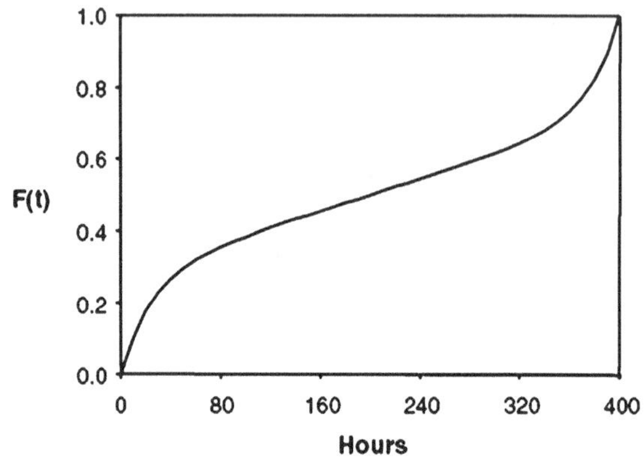

Fig. 2.7 Cumulative failure function [25]

At 20 h, an additional 72 joints fail and so forth. We can then plot the failure function as shown in Fig. 2.7, where the last joint failed at 400 h.

Then the failure probability density function (PDF) is:

$$f(t) = \frac{dF(t)}{dt} = -\frac{dR(t)}{dt}. \tag{2.7}$$

We can also define a complementary "unreliability" function $Q(t)$, given by:

$$Q(t) = 1 - R(t), \tag{2.8}$$

so from the failure PDF equation above:

$$f(t) = \frac{dQ(t)}{dt} = -\frac{dR(t)}{dt}, \tag{2.9}$$

and,

$$Q(t) = \int_0^t f(\tau)\, d\tau = F(t), \tag{2.10}$$

i.e., the "unreliability" $Q(t)$ is actually the cumulative failure function $F(t)$. Similarly:

$$R(t) = \int_t^\infty f(\tau)\, d\tau, \tag{2.11}$$

and,

$$\int_t^\infty f(\tau)\, d\tau = 1. \tag{2.12}$$

In many cases, the *hazard rate* $\lambda(t)$ (sometimes termed the *force of mortality*) is more useful, or more convenient to use, than the failure PDF $f(t)$. *The hazard rate is the failure rate normalized to the number of surviving operational parts*, i.e.,

$$\lambda(t) = \frac{1}{R(t)} \frac{dQ(t)}{dt} = \frac{-1}{R(t)} \frac{dR(t)}{dt}, \tag{2.13}$$

from which we can readily relate the hazard rate $\lambda(t)$ to the failure PDF $f(t)$ by:

$$\lambda(t) = \frac{f(t)}{R(t)}, \tag{2.14}$$

and determine the *cumulative hazard rate* $H(t)$, [analogous to the cumulative failure function $F(t)$], to be:

$$H(t) = \int_0^t \lambda(\tau)\, d\tau = -\ln R(t), \tag{2.15}$$

since $R(0) = 1$.

Another commonly used reliability criterion is the *mean time to failure* (*MTTF*), defined as:

$$MTTF = \int_0^\infty t \cdot f(t)\, dt, \tag{2.16}$$

which reduces to:

$$MTTF = \int_0^\infty R(t)\, dt, \tag{2.17}$$

after integration by parts, and provided that $R(t) \to 0$ for $t \ll \infty$.

These results and relationships are summarized for convenience in Table 2.2, which also includes additional information developed below. The example of a cumulative failure function shown in Fig. 2.8 corresponds to the bottom bathtub curve of Fig. 2.9.

2.7.1 Exponential Distribution

The simplest practical failure PDF, and one with wide application in electronics, is the single parameter *exponential distribution function*:

$$f_e(t) = \lambda_0 \cdot \exp(-\lambda_0 t) \cdot u(t), \tag{2.18}$$

Table 2.2 Reliability relationships, for $n_f(t)$ failed and $n_s(t)$ surviving at time t of an initial sample of n_0

Function	Definition	General	Two-parameter Weibull ($\gamma = 0$)	Exponential ($\eta = 1, \beta = 1$)	Rayleigh ($\beta = 2, k = 2/\eta^2$)
Probability density function $f(t)$	$\frac{1}{n_0}\frac{dn_f(t)}{dt}$	$\frac{dQ(t)}{dt} = -\frac{dR(t)}{dt}$	$\frac{\beta}{\eta}\left(\frac{t}{\eta}\right)^{\beta-1} \exp -\left(\frac{t}{\eta}\right)^{\beta}$	$\lambda_0 \cdot \exp(-\lambda_0 t)$	$kt \cdot \exp\left(-\frac{1}{2}kt^2\right)$
Survivor function (reliability) $R(t)$	$\frac{n_s(t)}{n_0}$	$\int_t^{\infty} f(\tau)d\tau$	$\exp -\left(\frac{t}{\eta}\right)^{\beta}$	$e^{-\lambda_0 t}$	$\exp\left(-\frac{1}{2}kt^2\right)$
Cumulative failure function (unreliability) $Q(t) = 1 - R(t) = F(t)$	$\frac{n_f(t)}{n_0}$	$\int_o^t f(\tau)\,d\tau$	$1 - \exp -\left(\frac{t-\gamma}{\eta}\right)^{\beta}$	$1 - e^{-\lambda_0 t}$	$1 - \exp\left(-\frac{1}{2}kt^2\right)$
Hazard rate $\lambda(t)$	$\frac{1}{n_s(t)}\frac{dn_f(t)}{dt}$	$-\frac{1}{R(t)}\frac{dR(t)}{dt} = \frac{f(t)}{R(t)}$	$\frac{\beta}{\eta}\left(\frac{t}{\eta}\right)^{\beta-1}$	λ_0	kt
Cumulative hazard rate $H(t)$	$\int_0^t \lambda(\tau)d\tau$	$-\ln R(t)$	$\left(\frac{t}{\eta}\right)^{\beta}$	$\lambda_0 t$	kt^2
Mean time to failure *MTTF*	$\int_0^{\infty} t \cdot f(t)\,dt$	$\int_0^{\infty} R(t)\,dt$	–	$1/\lambda_0$	–

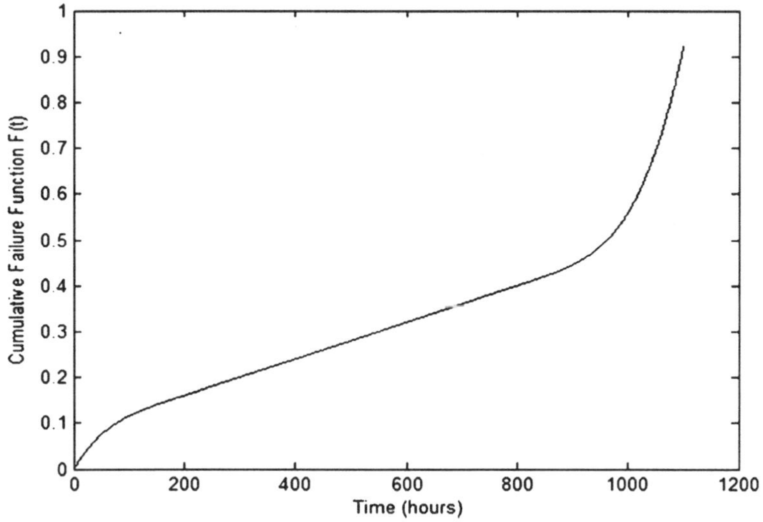

Fig. 2.8 Cumulative failure function

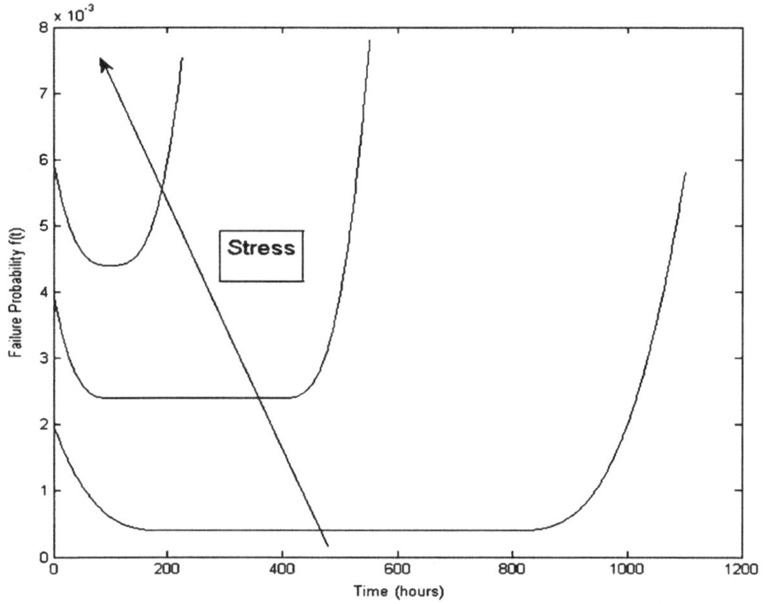

Fig. 2.9 Bathtub curve changes with increased stress (e.g., temperature, voltage, etc.)

where $u(t)$ is the Heaviside step function [$u(t) = 0$ for $t < 0$, and $u(t) = 1$ for $t > 0$] and λ_0 is called the *chance hazard rate*. Applying the formula above:

$$Q_e(t) = \int_0^t f_e(\tau)\, d\tau = \lambda_0 \int_0^t e^{-\lambda_0 t} dt = 1 - e^{-\lambda_0 t}, \tag{2.19}$$

$$R_e(t) = \int_t^{\infty} f_e(\tau)\,d\tau = \lambda_0 \int_t^{\infty} e^{-\lambda_0 t} dt = e^{-\lambda_0 t}. \tag{2.20}$$

Similarly, the hazard rate, $\lambda_e(t)$, and *MTTF* are given by:

$$\lambda_e(t) = -\frac{dR_e(t)/dt}{R_e(t)} = -\frac{-\lambda_0 e^{-\lambda_0 t}}{e^{-\lambda_0 t}} = \lambda_0, \tag{2.21}$$

and,

$$MTTF_e = \int_0^{\infty} R_e(t)dt = \frac{-1}{\lambda_0}[\exp(-\lambda_0 t)]_0^{\infty} = \frac{1}{\lambda_0}, \tag{2.22}$$

respectively, which suggest the interpretation of reliability as:

$$R_e(t) = e^{-1/MTTF_e}. \tag{2.23}$$

Note that the constant hazard rate result does not equate to the constant failure rate at the bottom of the bathtub curve, but there can be a functional approximate equivalence:

$$\lambda(t) = \frac{1}{n_s(t)} \frac{dn_f(t)}{dt} \approx f(t) = \frac{1}{n_0} \frac{dn_f(t)}{dt}, \tag{2.24}$$

where $n_f(t)$ have failed and $n_s(t)$ survived at time t of sample size n_0, if $n_s(t) \approx n_0$, i.e., if most of the original sample survives.

In plotting failure data to determine the specific parameter (or parameters) to characterize the specific component set, one counts the failures as a function of time, and the accumulated (total number) of failures, $Q(t)$, and uses the cumulative hazard rate, $H(t)$, in the form:

$$H_e(t) = \ln \frac{1}{R_e(t)} = \ln \frac{1}{1 - Q_e(t)} = \lambda t. \tag{2.25}$$

For the exponential distribution, with $Q_e(t) = 1 - \exp(-\lambda t)$, the slope of the log-linear plot of $1/(1 - Q(t))$ vs. t (which necessarily passes through the origin) gives $\lambda = MTTF^{-1}$ (Fig. 2.10), from:

$$\ln \frac{1}{1 - Q_e(t)} = \lambda t. \tag{2.26}$$

There is usually such an implicit assumption made of which mathematical distribution would apply in deciding how to plot the failure data (and hence implicitly of the failure mechanism).

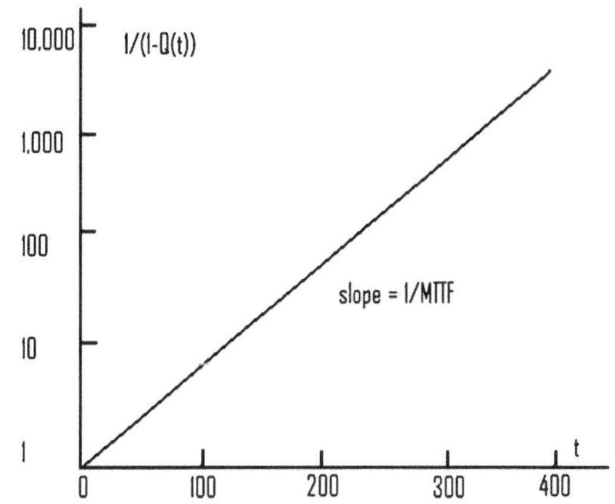

Fig. 2.10 Exponential distribution: Schematic semilog plot

2.7.2 *Weibull Distribution*

In fact, the exponential distribution is a special single parameter case of the more general three-parameter *Weibull distribution*:

$$f_w(t) = \frac{\beta}{\eta}\left(\frac{t-\gamma}{\eta}\right)^{\beta-1} \exp - \left(\frac{t-\gamma}{\eta}\right)^{\beta}, \tag{2.27}$$

which yields reliability:

$$R_w(t) = \exp - \left(\frac{t-\gamma}{\eta}\right)^{\beta}. \tag{2.28}$$

The effects of *shape parameter/factor* β, *scale parameter/factor* η, and *location parameter/factor* γ, are shown in Fig. 2.11, with the Weibull hazard rate and cumulative hazard rate variations with shape factor illustrated in Fig. 2.12.

For the shape factor ($\beta > 0$):

- $\beta < 1$ represents early failure (including burn-in)
- $\beta = 1$ represents constant failure rate (e.g., the bottom of the bathtub curve)
- $\beta > 1$ represents the wearout phase

For the location parameter, γ:

- $\gamma < 0$ covers a preexisting failure condition before $t = 0$, e.g., from storage, transport, etc.
- $\gamma = 0$ represents the usual condition where failure begins at $t = 0$
- $\gamma > 0$ applies when there is a failure-free period until $t = \gamma$ (Note that the symbol α is often used instead of η for the scale factor.)

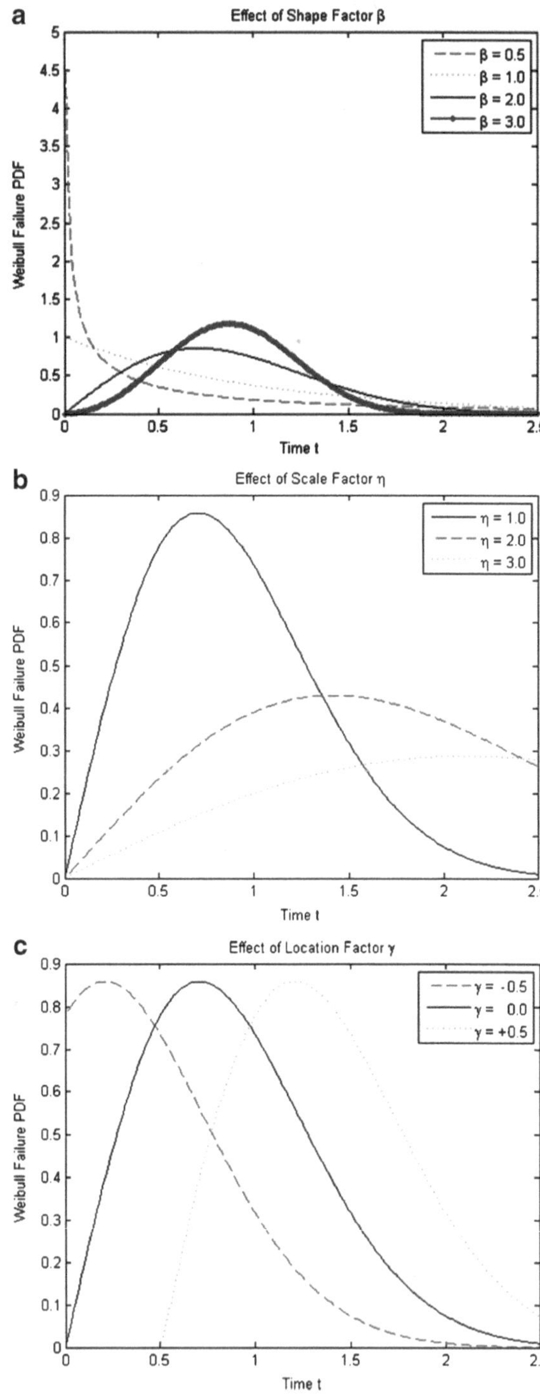

Fig. 2.11 Weibull distribution: failure probability density function (PDF). (**a**) Effect of shape parameter $\beta(\gamma = 0, \eta = 1)$; (**b**) Effect of scale parameter $\eta(\gamma = 0, \beta = 2)$; (**c**) Effect of location parameter $\gamma(\eta = 1, \beta = 2)$

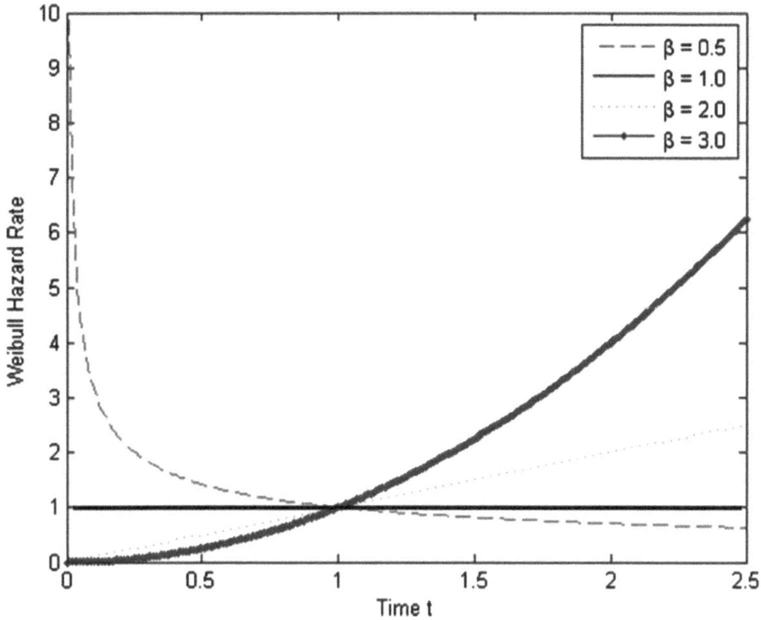

Fig. 2.12 Weibull distribution: Effect of shape factor β ($\gamma = 0, \eta = 1$)

For the usual case, where $\gamma = 0$, the three-parameter Weibull distribution reduces to the two-parameter Weibull distribution, which then reduces further to the exponential distribution above with $\beta = 1$ and $\eta = 1/\lambda_0$ (see Figs. 2.11a and 2.12); Note from Fig. 2.12 that $\beta = 0.5$, 1, and 3 can be seen as representative of the early failure, constant failure phase, and wearout sections of the bathtub curve, respectively (provided that $n_f < n_0$, as explained above).

Again, the cumulative hazard rate, for $Q_w(t) = 1 - \exp - (t/\eta)^\beta$, gives:

$$H_w(t) = \ln \frac{1}{1 - Q_w(t)} = \left(\frac{t}{\eta}\right)^\beta, \tag{2.29}$$

so, to find both β and η, one plots $\log\{\log[1/(1 - Q(t))]\}$ vs. $\log t$, and gets η from the intercept, and β from the slope (Fig. 2.13), as:

$$\ln \ln \frac{1}{1 - Q_w(t)} = \beta \ln t - \beta \ln \eta, \tag{2.30}$$

and Weibull log.log vs. log graph paper is available to facilitate direct plotting.

Table 2.2 includes a summary of the Weibull and exponential distributions, and of the Rayleigh distribution, which is specified by $\gamma = 0$, and $\beta = 2$, where $\eta = \sqrt{2/k}$.

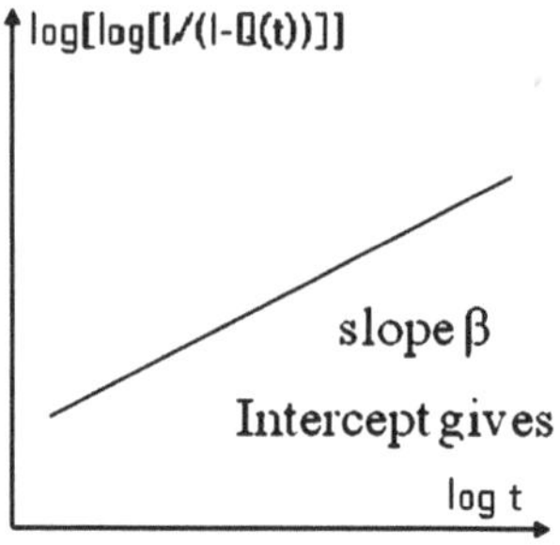

Fig. 2.13 Weibull distribution: Schematic plot

In the context of the location parameter, it is worth noting that the assignment of $t = 0$ is clearly arbitrary, and we can define a *conditional reliability*, $R(t, T)$, as the probability of the (nonrepairable) device or system to operate for time, t, having already operated for time, T, as:

$$R(t,T) = \frac{R(t+T)}{R(T)} = \frac{\exp -\lambda(t+T)}{\exp -\lambda T} = \exp -\lambda t. \tag{2.31}$$

2.7.3 Log-Normal Distribution

The other failure probability distribution with significance in microelectronics is the *log-normal*, given by:

$$Q_{LN}(t) = \int_0^t \frac{1}{t\sigma\sqrt{2\pi}} e^{-1/2((\ln t-\mu)/\sigma)} dt = \frac{1}{\sqrt{2\pi}} \int_{-\infty}^{(\ln t-\mu)/\sigma} e^{-1/2z^2} dz, \tag{2.32}$$

where $z = (\ln\, t - \mu)/\sigma$, $\mu = \ln\, T_{50}$, and T_{50} is defined as the time t_f at which $Q_{LN}(t_f) = R_{LN}(t_f) = 0.5$ or 50%. Re-writing the expression for z:

$$\ln t = \sigma z + \mu = \sigma\left[\Phi^{-1}(Q_{LN}(t))\right] + \ln T_{50}, \tag{2.33}$$

where $\Phi^{-1}(Q_{LN}(t))$ is a tabulated function of $Q_{LN}(t)$. This equation can be plotted readily on log-normal graph paper, where the log(time-to-failure) x-axis is labeled in terms of time, and the y-axis (which is actually $\Phi^{-1}(Q_{LN}(t))$) is labeled as the cumulative failed percentage (or failure probability), $Q(t)$. (Figure 2.14 shows such a plot, but with the axes reversed to match the form of the equation above. Since $z = 0$ at $t = T_{50}$, $Q = 50\%$, and $z = -1$ at $t = T_{15.9}$, $Q = 15.9\%$, the slope of this plot gives *slope* $= \sigma = \ln(T_{50}/T_{15.9})$.) The log-normal parameters $f_{LN}(t)$, $\lambda_{LN}(t)$, and $Q_{LN}(t)$, are shown for completeness as functions of time in Fig. 2.15, with $T_{50} = e^{\mu}$ *as the third parameter.*

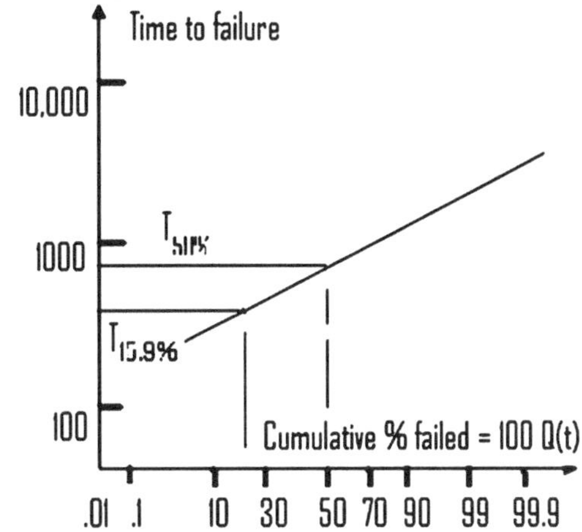

Fig. 2.14 Log-normal distribution: Schematic plot

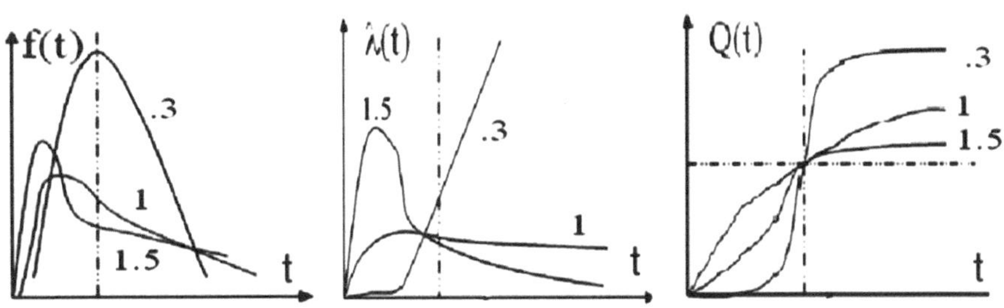

Fig. 2.15 Log-normal distribution: schematic plots of distribution parameters (failure PDF, hazard rate, and cumulative failures) vs. time

2.7.4 *Physical Basis of the Distributions*

For $\beta = 1$, the Weibull distribution reduces to the exponential, with $\lambda = \mathit{constant} = \lambda_0$, corresponding to the bottom of the bathtub curve. In modern electronic systems, the apparently constant hazard rate actually masks many different and unrelated failure modes with different temporal rates. In this case, any attempt to model these varied physical failure processes with a single parameter set, as if they were all due to the same failure mechanism, would clearly be flawed. In particular, if one then tried to apply a single formula for thermal variation of these many different physical processes to predict failure rates under other circumstances, the results would be unlikely to match reality. This was the fundamental problem when Military Handbook Standard 217, which was developed to model and predict single constant hazard rate failure modes, began to be applied to the apparently constant rate of modern complex systems, which result from low random failure rates from diverse causes. For the Rayleigh distribution, where $\beta = 2$, $\lambda(t)$ is directly proportional to t, and clearly this case corresponds to failure due to time-accumulated damage.

Where only one defect site causes the failures, and the damage susceptible population is removed, $\lambda(t)$ decreases as time goes on, and corresponds to the early failure (burn-in) section of the bathtub curve, e.g., for the Weibull distribution with $\beta = 0.5$.

A log-normal distribution can arise in the following way, if failure occurs when a resistance, r, increases to the failure threshold r_n. Assume that the resistance increases with each time step as $r_{i+1} = r_i(1 + \delta)$ *from the initial value* r_0, then:

$$r_n = r_0 \prod_{t=0}^{n} (1 + \delta_t), \tag{2.34}$$

$$\ln r_n = \ln r_0 + \sum_{i=0}^{n} \ln(1 + \delta_i) \approx \ln r_0 + \sum_{i=0}^{n} \delta_i, \tag{2.35}$$

where δ_i is an independently distributed random variable, and hence, log r_n is normally distributed and r_n (the end of life resistance) is log-normally distributed.

One might expect that at least the general nature of the physical origin of reliability failures might be determined by the fit or otherwise of the data to the distribution assumed for the graphical plots. However, the Weibull and log-normal plots of the same data shown in Fig. 2.16 demonstrate that the fit can be ambiguous.

2.8 A Generic Weibull Distribution Model to Predict Reliability of Microsystems

The classic two-parameter Weibull distribution has the following form:

$$F(t) = 1 - \exp\{-[(t)/\alpha]^{\beta}\}. \tag{2.36}$$

Then the failure intensity function is in this case:

$$f(t) = \frac{\beta}{\alpha}\left(\frac{t}{\alpha}\right)^{(\beta-1)} \exp\{-[(t)/\alpha]^{\beta}\}. \tag{2.37}$$

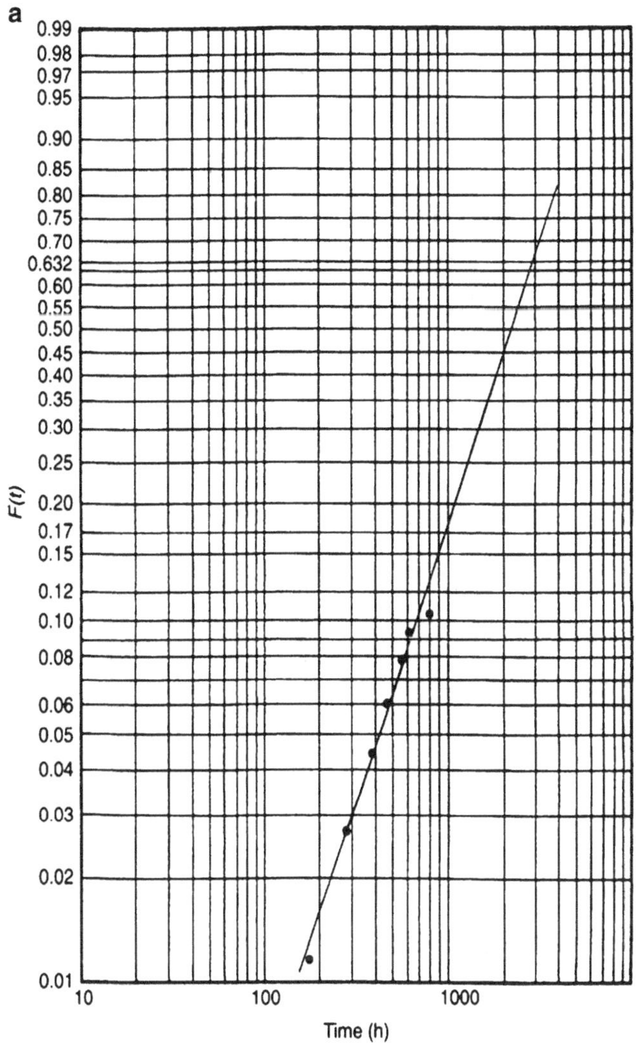

Fig. 2.16 Failure data plots. (**a**) Weibull probability paper (with permission [26])

The traditional three-parameter Weibull distribution has the following form:

$$F(t) = 1 - \exp\{-[(t - \tau)/\alpha]^{\beta}\}, t \geq \tau, \tag{2.38}$$

where α is the scale parameter, β is the shape parameter, which is a kind of wear characteristic or associated with different failure modes, and τ is the location parameter indicating the minimum life.

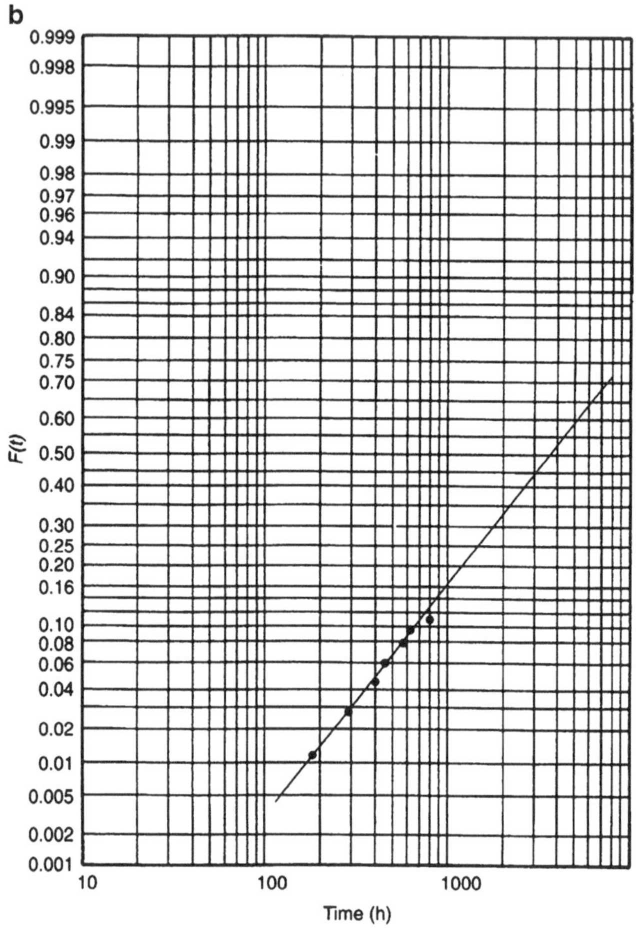

Fig. 2.16 (**b**) log-normal probability paper (with permission from [26])

2.8.1 Failure-Criteria Dependence of the Location Parameter

Since α and β in the Weibull distribution are material dependent with α characterizing the strength of the material and β characterizing the aging effect of the material, it can be assumed that they are independent of the failure criterion in this case. However, the location parameter τ should depend on the failure criterion because of the cumulative damage leading to a failure. A new 4-parameter Weibull distribution is described below. The new distribution is able to handle the resistivity change as a failure criterion. We demonstrate this by looking into the reliability of anisotropic conductive adhesive joints.

Let the failure criterion be generally described as $r > kr_0$, where r_0 is the nominal level.

Let τ_0 be the location parameter at the nominal value. Here we assume that this location parameter τ_0 is greater than 0, which means that the material has a failure-free life until τ_0. Some preliminary analysis indicates that a model for τ could be:

$$\tau = \tau_0 k^b, \tag{2.39}$$

where b is an empirical parameter to be estimated with test data. That is, the probability of failure at time t depends on the value of k in the following manner:

$$F(t;k) = 1 - \exp\{-[(t - \tau_0 k^b)/\alpha]^\beta\}. \tag{2.40}$$

Hence, this is a model with four parameters, but it can be fitted to the datasets under different criteria at the same time. Such a model is useful in many aspects. Some are discussed in the following.

First, the minimum life defined as $\tau = \tau_0 k^b$ can be computed for any given failure criterion. This provides a theoretical explanation of the existence of the minimum life and its dependence of the failure criteria.

Second, fixing a minimum cycle time to failure, the failure criterion that meets this requirement can be determined. This is useful in a contractual situation when a minimum cycle time is to be guaranteed. That is, if the required or guaranteed minimum cycle life is τ_r, from the inequality $\tau_0 k^b \geq \tau_r$, we get that $k \geq (\tau_r/\tau_0)^{1/b}$. In other words, the failure criteria can at least be set as $k = (\tau_r/\tau_0)^{1/b}$.

Furthermore, under any failure criterion, the cumulative failure probability can be computed at any time.

2.8.2 *Least Squares Estimation*

The general model contains four parameters that have to be estimated using the data from testing. Various methods can be used, and here the parameters can be estimated by a simple least square method. Here the cumulative distribution function is estimated by the mean ranking with the form:

$$\hat{F}_{ij} = \frac{j}{n_i + 1}, \tag{2.41}$$

and the sum of square deviation can be written as:

$$\begin{aligned} SSE &= \sum_{i=1}^{m}\sum_{j=1}^{d_i}[\hat{F}_{ij}(t) - F_{ij}(t)]^2 \\ &= \sum_{i=1}^{m}\sum_{j=1}^{d_i}\left[1 - \frac{j}{n_i+1} - \exp\{-[(t_{ij} - \tau_0 k_i^b)/\alpha]^\beta\}\right]^2, \end{aligned} \tag{2.42}$$

where m different failure criteria have been considered and n_i samples are tested for each criterion i and k_i is the criterion parameter of criterion i. For each sample group with n_i samples, d_i components have failed, and t_{ij} is the time to failure for the jth failed component.

Since parameter τ_0 is defined as the location parameter at the nominal failure criterion, it is suggested to use the sample data under the nominal failure criterion to estimate the parameter τ_0.

Therefore, the minimization of *SSE* is accomplished by taking the partial derivatives of SSE with respect to the parameters and setting the resulting equations to zero, which leads to:

$$\frac{\partial SSE}{\partial \alpha} = -2\frac{\beta}{\alpha}\sum_{i=1}^{m}\sum_{j=1}^{d_i}\left[1 - \frac{j}{n_i+1} - \exp\{-[(t_{ij} - \tau_0 k_i^b)/\alpha]^\beta\}\right] \cdot \exp\{-[(t_{ij} - \tau_0 k_i^b)/\alpha]^\beta\} \cdot [(t_{ij} - \tau_0 k_i^b)/\alpha]^\beta = 0, \tag{2.43}$$

$$\frac{\partial SSE}{\partial \beta} = 2\sum_{i=1}^{m}\sum_{j=1}^{d_i}\left[1 - \frac{j}{n_i+1} - \{-[(t_{ij} - \tau_0 k_i^b)/\alpha]^\beta\}\right] \cdot \exp\{-[(t_{ij} - \tau_0 k_i^b)/\alpha]^\beta\} \cdot (t_{ij} - \tau_0\, k_i^b)^\beta \ln[(t_{ij} - \tau_0\, k_i^b)/\alpha] = 0, \tag{2.44}$$

$$\frac{\partial SSE}{\partial b} = -2\tau_0\frac{\beta}{\alpha}\sum_{i=1}^{m}\sum_{j=1}^{d_i}\left[1 - \frac{j}{n_i+1} - \exp\{-[(t_{ij} - \tau_0 k_i^b)/\alpha]^\beta\}\right] \cdot \exp\{-[(t_{ij} - \tau_0 k_i^b)/\alpha]^\beta\} \cdot [(t_{ij} - \tau_0 k_i^b)/\alpha]^{\beta-1} k_i^b \ln k_i = 0, \tag{2.45}$$

and,

$$\frac{\partial SSE}{\partial \tau_0} = -2\frac{\beta}{\alpha}\sum_{i=1}^{m}\sum_{j=1}^{d_i}\left[1 - \frac{j}{n_i+1} - \exp\{-[(t_{ij} - \tau_0 k_i^b)/\alpha]^\beta\}\right] \cdot \exp\{-[(t_{ij} - \tau_0 k_i^b)/\alpha]^\beta\} \cdot [(t_{ij} - \tau_0 k_i^b)/\alpha]^{\beta-1} k_i^b = 0. \tag{2.46}$$

The above equations can be solved using a computer spreadsheet or software. Also note that the location parameter model $\tau_i = \tau_0 k^b$ should satisfy the condition that $\tau_i \leq \tau_{i1}$ under each failure criterion i.

2.8.3 *The Experiment and Data*

The general approach above is verified for the analysis of conductive adhesive joining in flip-chip packaging.

A significant number of accelerated reliability tests under well-controlled conditions based on single joint resistance measurement to generate significant reliability data for using anisotropic conductive adhesive (ACA) flip-chip technology on FR-4 substrate have been generated in the literature [26]. Nine types of ACA and one nonconductive film (NCF) were used. In total, nearly one thousand single joints were subjected to reliability tests in terms of temperature cycling between −40 and 125°C with a dwell time of 15 min and a ramp rate of 110°C/min. The chip used for this reliability test had a pitch of 100 μm. Therefore, the test was particularly focused on evaluation of the reliability of ultrafine pitch flip-chip interconnections using ACA on a low-cost substrate.

The reliability was characterized by single contact resistance measured using the four-probe method during temperature cycling testing up to 3,000 cycles. The failure definitions are defined as 20% increase, larger than 50 mΩ, and larger than 100 mΩ, respectively, using the in situ electrical resistance measurement technique. Usually when tests are carried out in different conditions or when the data are from different failure criteria, the datasets are analyzed separately. This usually involves a large number of combined model parameters, and there is no clear relationship between the model parameters.

The test setup: To study the reliability of conductive adhesive joints, contact resistance of single joints is one of the most important parameters. Therefore, a test chip was designed for four-probe measurement of single joints. The configuration of the test chip contains 18 single joints and two daisy chains (18 joints for each). The pitch of the test chips is 100 μm. Bump metallization of the chips is electroless nickel and gold. Table 2.3 summarizes some characteristic parameters of the test chip. Here the reliability study focused on the reliability of ACA joining, i.e., the characteristics of ACA joints together with the usage environment. A temperature cycling test was applied for the evaluation. The reliability of ACA joints was characterized by the change of contact resistance in the cycled temperatures. A total of 954 joints (53 chips) with different ACA materials were tested. Two chips with 36 joints were measured in situ with the four-probe method during testing up to 3,000 cycles, and other joints were taken out from the equipment every several hundred cycles to manually measure the resistance change in room temperature.

Most ACA joints were manually measured every several hundred cycles because of the capacity of the cabinet. A total of 918 joints (51 chips) were tested. Some of them, 126 joints of 7 chips, failed after only 200 cycles due to bad alignment, so they were screened out. The remaining 792 joints (44 chips) were tested for 1,000 cycles. Cumulative failures of the ACA flip-chip joints were measured manually at room temperature. According to different criteria (i.e., the resistance increase was over 20%, contact resistance was over 50 and 100 mΩ).

Test results and discussions: Cumulative fails of the in situ testing are shown in Fig. 2.17. The number of fails is dependent on the definition of the failure. Figure 2.18 shows three statistics on the cumulative fails respectively based on the different criteria: >20% of contact resistance increase; >50 mΩ; >100 mΩ. When the criterion was defined at 20% of resistance increase, after 2,000 cycles all joints had failed. This definition might be too harsh for those joints only having a contact resistance of several mΩ. The 20% increase means only a few milliohms is allowed to vary. In some case, the limitation is still within the margin of error of the measurement.

Table 2.3 Technical data of silicon test chips

Chip size (mm)	Bump size (μm)	Bump height (μm)	Pitch (μm)	Number of bumps
3.0 × 3.0	60	20	100	54

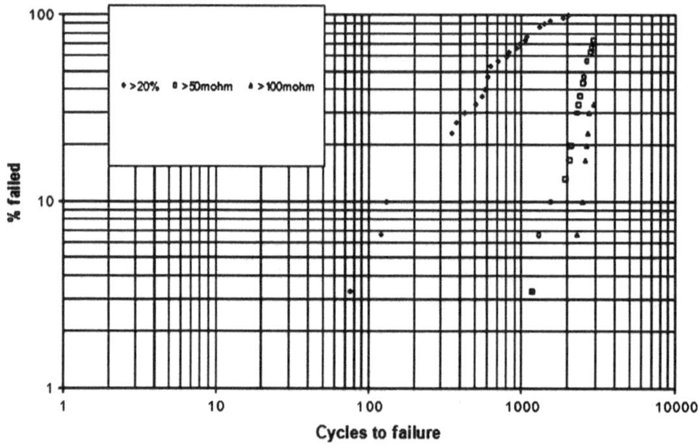

Fig. 2.17 Cumulative failure plot during temperature cycling test

Fig. 2.18 A typical trend of resistance change as a function of the number of cycles

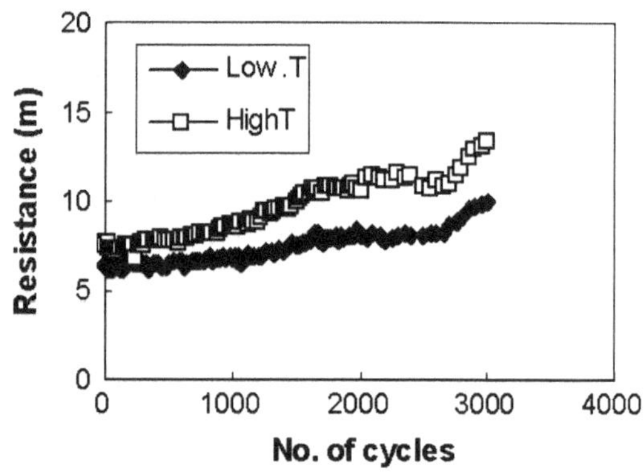

If we, in any case, allow 50 or 100 mΩ as the failure criteria, we will obtain a MTTF value of 2,500 and 3,500 cycles respectively from a simple Weibull probability plot. Therefore, it is reasonable that the criterion is defined according to the production requirements.

A problem in the analysis of this type of data is that failures under different criteria are usually analyzed separately. With a small number of data points and a large total number of model parameters, the analysis is usually inaccurate. It would be useful to develop an approach for joint analysis of the datasets. The following sections present a Weibull model with the analysis of the data in Fig. 2.18 as an example.

2.8.4 Analysis and the Results

Here we will follow the general model presented earlier. Data of the failure of adhesive flip-chip joints on an FR-4 substrate during the temperature cycling test

Table 2.4 Cumulative number of failures under different criteria

	Cumulative number of failures	
Cycles to failure	Criterion II (>50 m)	Criterion III (>100 m)
1,170	1	1
1,300	2	–
1,550	3	–
1,925	4	–
2,050	5	–
2,100	6	–
2,300	9	2
2,350	10	–
2,400	11	–
2,500	13	3
2,550	14	–
2,600	–	5
2,650	17	6
2,700	–	7
2,750	–	9
2,800	19	–
2,850	20	–
2,900	21	–
2,950	22	10

are considered to illustrate the above new model and estimation. Under criterion II (failure if resistance >50 mΩ) and criterion III (Table 2.4) (failure if resistance >100 mΩ), the cumulative number of failures are summarized in the following table: The nominal level of the test (r_0) is 6 mΩ, thus the criterion parameters k_1 and k_2 are $r_1/r_0 = 50/6 = 8.33$ and $r_1/r_0 = 100/6 = 16.67$ respectively. Using simple spreadsheet, the traditional least square estimation of the parameters is given by:

$$\alpha = 1{,}954,\ \beta = 4.076,\ \tau_0 = 370,\ b = 0.409 \text{ and } SSE = 0.1261.$$

The overall model is then given by:

$$F(t;k) = 1 - \exp\{-[(t - 370k^{0.409})/1954]^{4.076}\}, \tag{2.47}$$

where k is the failure criterion in terms of "failure when the resistance is k times the nominal value." This above formula can be used for different failure criteria.

2.8.5 Application of the Results

From the above analysis, note that the minimum cycle life is given by: Minimum life $= 370k^{0.409}$. Hence, for any given failure criterion, we can obtain the minimum life with this formula. The estimated minimum life at failure definition of larger than 50 and 100 mΩ are respectively 880.68 and 1169.33. Table 2.5 shows the estimated minimum life and MTTF under some different failure conditions.

Table 2.5 Minimum life (c_0) and MTTF under different criteria

k_0	c_0	MTTF
1	370.00	2143.082
2	491.34	2264.424
3	580.02	2353.100
4	652.48	2425.561
5	714.86	2487.944
6	770.23	2543.317
7	820.39	2593.468
8	866.46	2639.542
9	909.24	2682.326
10	949.30	2722.384
20	1260.63	3033.709
30	1488.14	3261.221
40	1674.05	3447.134
50	1834.11	3607.189

Furthermore, for fixed or agreed cycle to failure, we can obtain the maximum failure criteria as:

$$370{k_0}^{0.409} > c_0,$$

that is:

$$k_0 > \left(\frac{c_0}{370}\right)^{1/0.409}. \tag{2.48}$$

That is, to be sure that the minimum life is c_0, the failure criteria cannot be more stringent that "failure when the resistance is k_0 times the nominal value." This is, for example, a statement that can be used together with the minimum life requirement and can be added in the contractual situation.

Exercises

2.1 Explain the curve shown below. What is the reason for failures in the different regions?

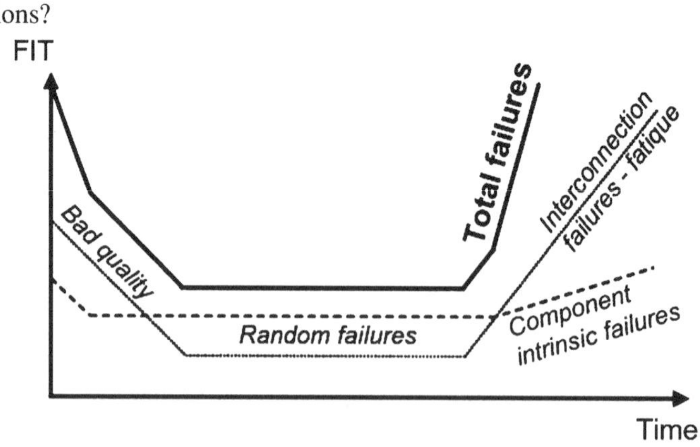

2.2 Consider the following equation:

$$R_{O/M} = 1 - \sum_{i=0}^{O-1} \left\{ \frac{M!}{(M-i)!i!} \right\} R^{i}_{\text{interconnection}} (1 - R_{\text{interconnection}})^{M-i}.$$

What is the solder joint reliability if seven of ten components work to maintain 90% system reliability? Plot the whole reliability diagram.

2.3 Draw the curve of the following equation that shows the failure rate function.

$$f(t) = \frac{\beta}{\alpha}\left(\frac{t}{\alpha}\right)^{\alpha-1} \exp\{-[(t)/\alpha]^{\beta}\}.$$

What conclusions can you draw from the curve?

2.4 The infant mortality and wear out sections of Fig. 2.9 are plotted for the sake of the example as simple cubic functions for t < 200 and t > 800 hours in the bottom curve. (a) Predict when the last component will die. (b) Verify that your result seems sensible by extrapolating the wear out section of Fig. 2.8.

2.5 Tables A and B below present failure time data for two sequences of lab experiments, each on 20 samples. (a) Plot the data for both Weibull and log-normal distributions. (b) Determine η and β for the Weibull plot, and σ for the log-normal plot. (c) What is the mean failure time for each? (d) Can you establish the appropriate reliability distribution model from the plots? (e) Compare the data sets, and discuss, noting the regular failure rate in Table A.

Table A

Failure	1	2	3	4	5	6	7	8	9	10
Time(hr)	40	98	150	210	255	295	351	402	445	503
Failure	11	12	13	14	15	16	17	18	19	20
Time(hr)	551	595	648	705	750	790	860	900	950	1008

Table B

Failure	1	2	3	4	5	6	7	8	9	10
Time(hr)	15	35	50	70	90	115	135	160	190	220
Failure	11	12	13	14	15	16	17	18	19	20
Time(hr)	255	290	335	380	440	510	600	730	950	1500

References

1. F. Jensen, "Electronic Component Reliability, Fundamentals, Modeling, Evaluation, and Assurance", John Wiley & Sons, New York, 1995.
2. Military Standard MIL-STD-721C, Definitions for Terms for Reliability and Maintainability, June, 1981.
3. International Standard, Information technology – Software product quality, Part 1: Quality Model, ISO/IEC FDIS 9126-1, ISO/IEC 2000.
4. MIL-HDBK-217, Reliability Prediction of Electronic Equipment, Version E, October, 1986, 166.

5. S. Arrhenius, "On the Reaction Velocity of the Inversion of Cane Sugar by Acids," Zeitschrift für Physikalische Chemie 4, 1889, 226ff.
6. D. L. Crook, "Evolution of VLSI Reliability Engineering", Proc. Int. Reliability Physics Symposium, 1990, 2–11.
7. W. Perry, "Specifications and Standard – A New Way of Doing Business", US Department of Defense Policy Memorandum, June, 1994.
8. TS-TSY-000332, Reliability Prediction Procedure for Electronic Equipment, Issue 2, Bellcore, July, 1988.
9. Standard Reliability Table for Semiconductor Devices, Nippon Telegraph and Telephone Corporation, March, 1985.
10. Handbook of Reliability Data for Components Used in Telecommunications Systems, Issue 4, British Telecom, January, 1987.
11. Recuiel De Donnes De Fiabilite Du CNET (Collection of Reliability Data from CNET), Centre National D'Etudes des Telecommunications (National Center for Telecommunication Studies), 1983.
12. SN29500, Reliability and Quality Specification Failure Rates of Components, Siemens Standard, 1986.
13. J. B. Bowles, "A Survey of Reliability-Prediction Procedures for Microelectronic Devices", IEEE Transactions on Reliability, 41, 1992, 2–12.
14. CALCE. http://www.calce.umd.edu/
15. N. Kelkar, A. Fowler, M. Pecht and M. Cooper, "Phenomenological Reliability Modeling of Plastic Encapsulated Microcircuits", International Journal of Microcircuits and Electronic Packaging, 19(1), 1996, 167.
16. T. Stadterman, B. Hum, D. Barker and A. Dasgupta, "A Physics-of-Failure Approach to Accelerated Life Testing of Electronic Equipment", Proc. for the Test Technology Symposium '96, US Army Test and Evaluation Command, John Hopkins University/Applied Physics Laboratory, 4–6, June, 1996.
17. J. R. Black, "Physics of Electromigration", Proc. International Reliability Physics Symposium, 1974, 142–164.
18. J. Liu, L. Cao, M. Xie, T. N. Goh and Y. Tang, "A General Weibull Model for Reliability Analysis Under Different Failure Criteria – Application on Anisotropic Conductive Adhesive Joining Technology", IEEE Component Packaging and Manufacturing Technology, 28(4), October 2005, 322–327.
19. J. P. Clech, "Solder Reliability Solutions, A PC-Based Design-For-Reliability Tool", Proc. of Surface Mount International, SMTA, San Jose, CA, 1996.
20. A. Syed, "Predicting Solder Joint Reliability for Thermal, Power and Bend Cycle within 26% Accuracy", Proc. of the 51st Electronics Components and Technology Conference, Orlando, 2001, 255–263.
21. J. Galloway, L. Li, R. Dunne and H. Tsubaki, "Analysis of Acceleration Factors Used to Predict BGA Solder Joint Field Life", Proc. SMTA International, Chicago, 2001, 357–363.
22. M. S. Moosa and K. F. Poole, "Simulating IC Reliability with Emphasis on Process-Flaw Related Early Failures", IEEE Transactions on Reliability, 44, 1995, 556–561.
23. E. Demko, "Commercial-Off-The-Shelf (COTS): A Challenge to Military Equipment Reliability", Proc. Annual Reliability and Maintainability Symposium, Las Vegas, NV, 1996, 7–12.
24. T. Ejim, "High Reliability Telecommunications Equipment: A Tall Order for Chip-Scale Packages", Chip Scale Review, 5, 1998, 44–48.
25. R. Tummala, "Fundamentals of Microsystems Packaging", McGraw-Hill Professional, New York, 2001.
26. M. Ohring, "The Mathematics of Failure and Reliability" in "Reliability and Failure of Electronic Materials and Devices", Academic Press, San Diego, CA, 1998, 200–201, Chapter 4.

Chapter 3
General Failure Mechanisms of Microsystems

Abstract In this chapter, general failure mechanisms of microsystems are presented. First, the definition of failure is shortly discussed, followed by revealing of some of the most common failure mechanisms related to electronic components and their interconnections. Also the influences of the failure modes are shortly discussed as well as the failure preventive actions.

The failure modes are categorized in certain subclasses based on their physical behavior and the nature of the initial stress conditions. For instance, a rapid failure caused by overstress environment can be separated from the failure mechanisms that initiate and evolve slowly in wearout conditions. To prevent or at least delay the failures to occur, one should be aware of the basic failure mechanisms involved to electronics assemblies, or more precisely to electronics microsystems.

3.1 Introduction

The definition of failure can be phrased as "any condition that causes a device or circuit to fail to operate in a proper manner" [1]. A failure can occur instantly after a shock or develop slowly and degrade the product functionality over time and finally result as final failure. Failure causes of an electronic system can generally be divided into three categories: hardware, software, and design-related failures, Table 3.1 [2]. According to a study made by RiAC, hardware-related failures account for 58% of the total failures, of which approximately half is allocated to components. The rest of the hardware failures are related to the manufacturing and the usage environment. Software and design immaturity represent a 22% fraction of the failures. The rest of the customer-returned products were fully functional.

In material sciences, a failure is defined as gradual material property deterioration resulting from subsequent events under stress conditions. For example, fatigue crack starts from initiation, continues with propagation, and ends finally to full rupture. This is an example of a failure mechanism. There are numbers of other hardware- and software-related failure mechanisms involved. These failure mechanisms are either systematic or random in their nature. An example of

J. Liu et al., *Reliability of Microtechnology: Interconnects, Devices and Systems*,
DOI 10.1007/978-1-4419-5760-3_3,

Table 3.1 Failure causes distribution of an electronic system [2]

Category	Failure cause	Description/examples	Fraction (%)
Hardware	Parts	ICs, transistors, resistors, connectors, etc.	22%
	Manufacturing	Anomalies in the manufacturing process, i.e., solder joint defects, etc.	15%
	Wearout	Component-related examples are drying electrolytic capacitors and switch wearout	9%
	Induced	External applied stress, i.e., dropping, bending, electricity, etc.	12%
Software	Software	Failures of a system to perform its intended function due to the manifestation of software fault	9%
Design	Design	Failures resulting from an inadequate design, i.e., tolerance stack-up, unanticipated logic conditions	9%
	System management	Failures related to faulty interpretation of system requirements or errors in the subpart interfaces, etc.	4%
No failure	No defect	Perceived failures that cannot be reproduced upon further testing	20%

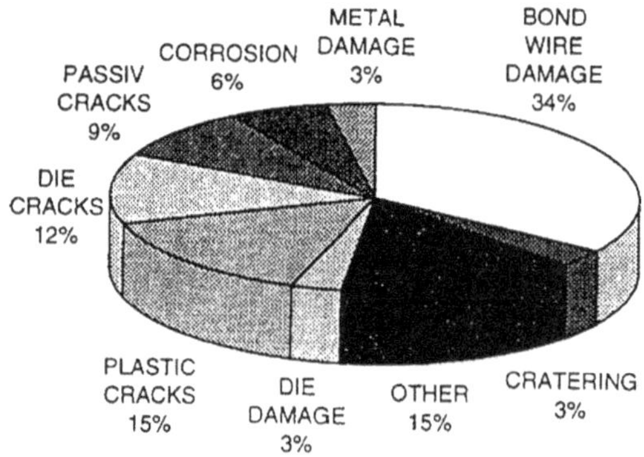

Fig. 3.1 Example of failure causes of component with ICs

component failure causes is presented in Fig. 3.1. This chapter discusses some of the common electronics material system-related failure mechanisms.

In the assessment of reliability performance of a microstructure, the stress distribution, the strain amplitude, the strain rate, the cyclic nature of the stress (either mechanical or thermal), the temperature, and other environmental factors all have to be considered. Basic processes or factors that are believed to be probable reasons for solder failure while in service are inferior or of inadequate mechanical strength, creep, mechanical and/or thermal fatigue, thermal expansion anisotropy, corrosion-enhanced fatigue, intermetallic compound formation, detrimental microstructure development, voids, electromigration, and leaching.

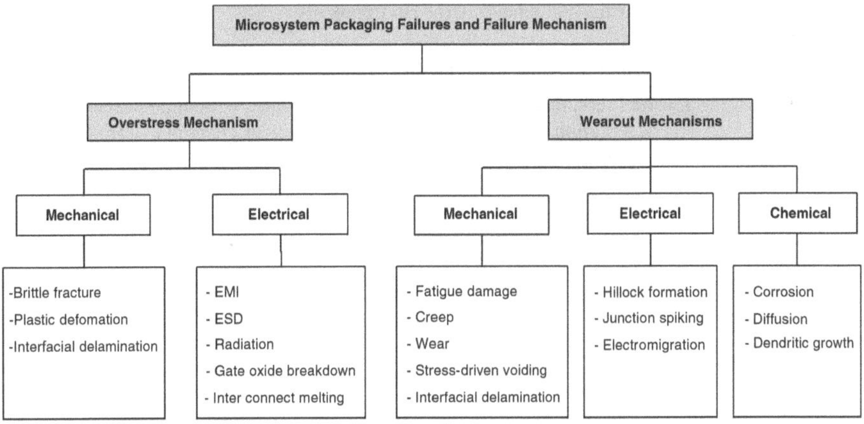

Fig. 3.2 Overview of the failure mechanisms in microsystems [3]

A particular failure mode is the result of a certain failure mechanism in which certain specific combinations of material properties and the surrounding environment act together. There are three major mechanisms of an interconnect failure, namely, tensile rupture (fracture due to mechanical overloading), creep failure (damage caused by a long-lasting permanent load or stress), and fatigue (damage caused by cyclical loads or stresses). These three mechanisms often interplay simultaneously with each other. Most of the physical failure modes (fatigue, delamination, creep, etc.) are generally a result of thermomechanical stresses. There are, however, other factors that also affect the failure behavior of microsystems. Electrical and chemical actions can also be responsible for many thermomechanical failures in modern electronic packages. Electromigration-induced voiding is, for example, primarily due to high electrical current density. Corrosion is another factor that accelerates fatigue and delamination failure.

A microsystem alone cannot be considered as reliable or unreliable according to the definition of reliability above; it becomes so only in the context of a microsystem where the components are connected via interconnects to the substrate. The properties of the component, substrate, and interconnects, together with service conditions, design life span, and the acceptable failure probability for the whole assembly determine the reliability of the microsystem. A general overview of the failure mechanisms in microsystems is shown in Fig. 3.2.

3.2 Mechanical and Thermomechanical Failure Mechanisms

Mechanical and thermomechanical degradation mechanisms can generally be categorized to (1) fatigue, (2) creep, (3) corrosion, (4) brittle fracture, Table 3.2. The creep and fatigue are typical solder joint system wearout failure mechanisms, which slowly evolve and result after a period of time as failure. Corrosion is also a

Table 3.2 Failure mechanisms and related stress environments

Failure mechanism	Stress environment	Description
Fatigue [4]	Cyclic stress, thermal [4] or vibration [5]	Fatigue failure initiates with microcrack and propagates to solder crack in cyclic stresses [5]
Creep [4]	Long-lasting permanent load [4]	Global plastic deformation of solder under static mechanical stress and temperature [6]
Corrosion [7]	Galvanic pair [7]	Metals with different electrochemical potential are in contact to cause material loss in the anode metal [7]
Brittle fracture [8]	Drop or shock [8]	Fracture occurs in the brittle intermetallics layer of solder joint [8]

relatively slow failure mechanism, which can result in visual deterioration or to failures in signal paths or to degraded mechanical structures. Brittle fractures are typically limited to incompatible materials in excess stress conditions, i.e., in drop conditions.

3.2.1 Low Cycle Fatigue

When component interconnections experience cyclic temperature changes, stresses will be induced to the solder–component and solder–substrate interfaces. This is due to uneven expansions of the different materials involved, Fig. 3.3. The stresses will cause deformation in the solder joint structure. With low stress level, the solder deformation will be reversible, when the applied stress is finished. Although the stress level is below the yield strength of the solder material, a long-lasting cyclic stress can lead to irreversible plastic deformation. This is called as low cycle fatigue.

The fatigue failure mechanism can be divided into three phases: (1) crack nucleation, (2) crack propagation, and (3) final fracture; Fig 3.4. The crack nucleation is preceded by microstructural changes, e.g., local grain growth [5]. The crack propagation starts with crystallographic propagation, which is followed by non-crystallographic propagation [9]. After enough plastic deformation has taken place, the mechanical strength and electrical path integrity of the solder joint has totally gone, meaning that final fracture has been fully developed, Fig. 3.5. The solder fatigue failures typically occur suddenly and unexpectedly as no observable plastic deformation occurs before the failure.

Under typical usage conditions, the oxygen is absorbed into the metal surfaces. Furthermore, the metal surface reacts with oxygen, which produces metal oxides. When the metal object is cyclically stressed, the chemisorbed oxygen is dissolved into the metal and close to the slip planes during the tension/compression cycles. Oxygen plays an active role in assisting early crack growth, but is not the cause of crack initiation. Smaller grain size, higher temperature, and lower stress tend to mitigate crack initiation. Larger grain size, lower temperatures, and higher stress

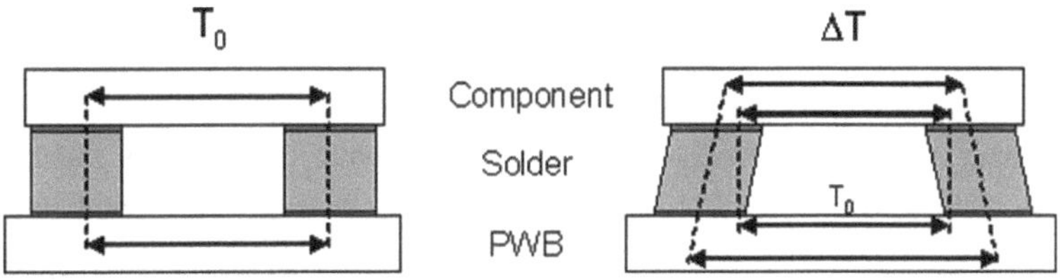

Fig. 3.3 Schematic illustration of component interconnections in temperature cycling

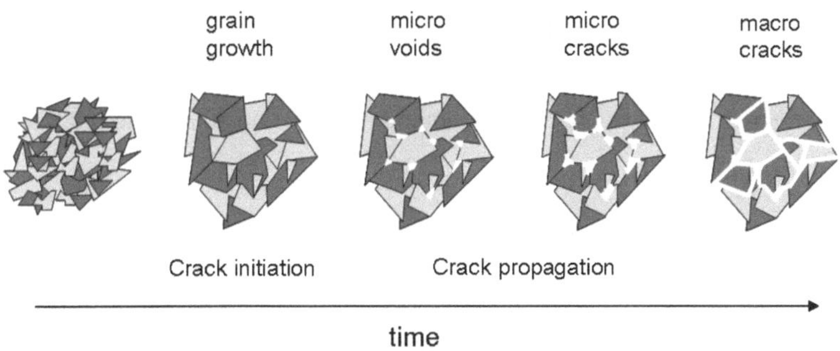

Fig. 3.4 Schematic illustration of the development of fatigue failure

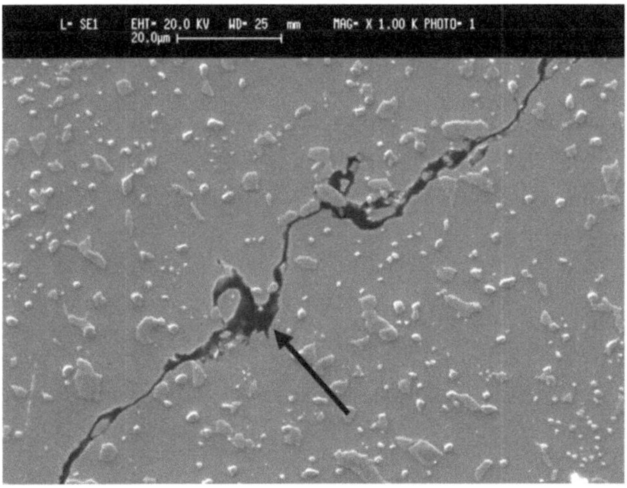

Fig. 3.5 Cross section of thermally cycled SnAgCu solder joint, taken with SEM. *Arrow* points to fatigue crack, which has fully developed. The mechanical properties and electrical signal path integrity have totally gone

tend to favor crack propagation. Dislocations, which are flaws in the crystal lattice, play an important role in the initiation of a crack. During fatigue they cluster and form first a channel-vein and later a ladder-like structure found in persistent slip bands (PSBs). The PSB emerges at the free surface as an extrusion of a material. Strain localization occurs when the dislocation pattern becomes locally unstable at a critical stress or strain and thin lamellae of PSBs are formed [10].

The main factors that affect of fatigue life are (1) microstructure of the material (grain size and texture), (2) processing (deformation history and manufacture), (3) load spectrum (sign, magnitude, rate, and history), (4) environment (temperature and corrosive medium), and (5) geometry of component (surface finish, notches, welds, connections, and thickness) [10].

3.2.2 Creep

When the solder joint is put under permanent, long-lasting static load, global plastic deformation will occur. This is known as creep, which is a measure of strain as a function of time [4]. Creep exists already at relatively low stresses. This implicates that solder strength is much lower with long-term loading than with short-term loading [11]. Creep rates are material and alloy related, for instance the creep rates of lead-free 96.5Sn3.5Ag solder joints are higher than those of 63Sn37Pb solder [12].

In principle, the metal deformation under static load can be divided into multiple phases, called primary creep, secondary creep, and tertiary creep, Fig. 3.6. At the primary creep strain stage, any microstructural evidence of creep damage can be found from the material. The secondary creep is also known as steady-state creep, as the strain level will maintain relatively constant. At this stage, the work hardening rate is balanced by thermally activated recovery rate. Individual voids start to occur at the microstructure level. At the tertiary creep region, the material experience higher strain rates than at the secondary creep. At this stage, the voids start to grow and to form cracks that will end up in final rupture.

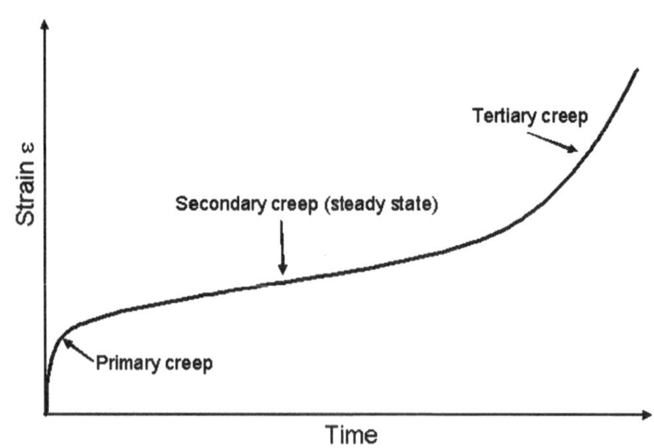

Fig. 3.6 Strain vs. time under constant load

The secondary creep mechanism plays a major role in the creep-fatigue damage in solders [13]. Secondary creep is controlled, for example, by the movement of dislocations in the slip planes [14]. Creep in solder is due to dislocation climb mechanism or due to grain boundary sliding and by intergranular or transgranular void migration (grain boundary diffusion) [15]. Most relevant creep deformation mechanisms with lead-free solders are dislocation creep, diffusion creep, and grain boundary sliding [16]. In recent years, many lead-free creep constitutive models have been developed, which emphasizes the importance of understanding the creep mechanisms behind the solder joint failures [17–19].

3.3 Brittle Fracture

Brittle fracture does not evolve as a result of plastic deformation, but is an instant rupture through material or several materials due to excess external stress. Breaking glass is a good example of such a failure event, in that the failure could easily be detected but no evidence when it was going to happen could have been recorded. In electronics, brittle factures occur typically in shock loading environments, for example, as a result of drop, where over 1,000 Gs can be generated. Brittle fracture is typically due to incompatible material selections.

Brittle fractures have typically been addressed in ceramic materials (e.g., LTCC or alumina) or brittle intermetallics (e.g., $(Cu,Ni)_6Sn_5$ or $AuSn_4$) in solder–pad interfaces. With ceramic components, brittle fractures can occur as (1) lifting or peeling of the pad, (2) rupture in ceramic material, or (3) fractured solder joint [20]. In the case of brittle intermetallics, the following failure modes have been reported: (1) fracture in the $(Cu,Ni)_6Sn_5$ layer, (2) fracture in the Cu_6Sn_5 layer on the Cu-OSP pads, or (3) fracture below the $(Cu,Ni)_6Sn_5$ layer on the Ni(P)|Au assemblies [8].

3.4 IC Level Failure Mechanisms

There are multiple failure mechanisms at the IC level. These include electromigration, dielectric breakdown, ionic contamination, dipole polarization, hot-carrier effects, and many others [21]. Some sources for the failures are listed below:

- Contamination from production (fingerprints, etc.) and use (corrosive gases, etc.)
- Electrical discharge
- Mechanical Shock and vibration
- Temperature (steady-state, ranges, gradients, and number of cycles)
- Humidity
- Pressure
- Radiation

The next sections discuss two of the important and typical failure mechanisms that can be avoided by proper designing or handling.

3.4.1 Electromigration

Electromigration refers to the migration or displacement of metal atoms due to the impact of moving electrons, Fig. 3.7. Electromigration differs from the normal diffusion, where concentration gradient is the driving force for the motion of atoms. With a high electrical current density, the atoms of the conductor trace move in the direction of the electrons toward the positive electrode (Fig. 3.7). This causes voids, or vacancies, in the negative electrode and decreases the cross-sectional area of the conductor. The current density is then further increased and electromigration is accelerated. Eventually, the conductor will have opens and hillocks in it. Factors that play a major role in electromigration are (1) temperature and its gradient, (2) current density, (3) conductor dimensions, (4) conductor grain structure, and (5) impurities [21].

An empirical model, based on J.R. Black's work from the late 1960s, was developed to estimate the mean time to failure (MTTF) caused by electromigration in IC structures. The MTTF (in hours) is calculated from the equation:

$$MTTF = AJ^{-n}e^{Q/kT}, \tag{3.1}$$

where A is a material-related parameter (hmA^2/μm^4), J is the current density involved (mA/μm^2), n represents a modeling exponent, Q is the activation energy, k is Boltzmann's constant (8.6×10^{-5} eV/K), and T is the temperature in Kelvins [22]. MTTFs of an IC with three different current densities as a function of temperature are presented in Fig. 3.8. As can be seen, with higher temperatures or current densities the expected MTTF is shorter.

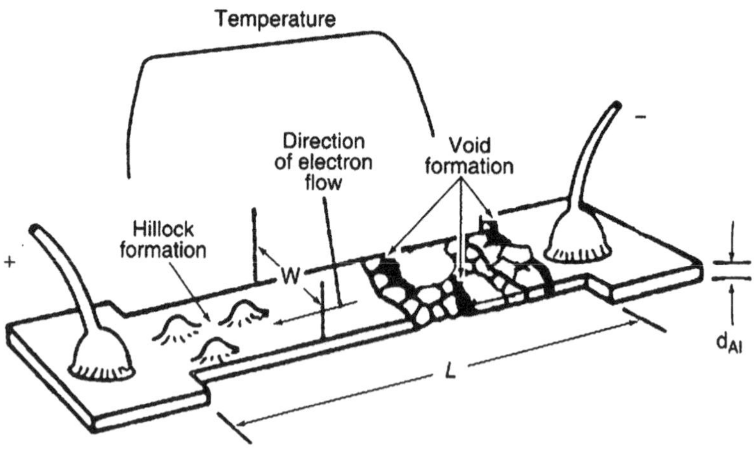

Fig. 3.7 Schematic illustration of damage in metallization trace caused by electromigration

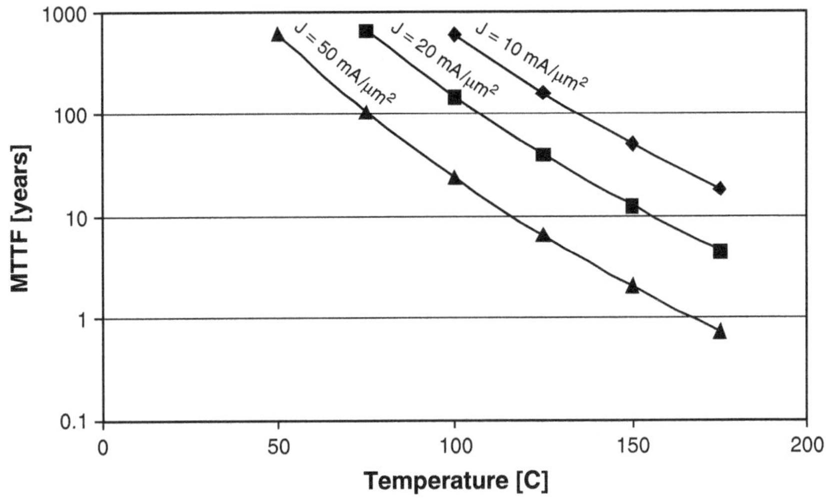

Fig. 3.8 MTTF of an IC due to electromigration with three current density levels (J). Example is calculated by using (3.1) with the activation energy of 0.67 eV, material constant of 0.44 hmA^2/µm^4, and modeling exponent of 2

3.4.2 *Electrostatic Discharge*

When the surfaces of two materials closely glide against each other, they are electrically charged. Depending on the molecular structure of the materials, there is a tendency for one surface to strip electrons from the other. This is commonly called as triboelectric charging [23]. An everyday example of the triboelectric charging is when person walks across a carpet and is then electrically charged up to ca. 20 kV. When person touches an object that is at different electrical charging level, the charge will be leveled between the person and the object. The discharging takes place within less than 100 ns. If the discharge path includes weak parts, e.g., material layers in nanometer scale, etc., the relatively high current density will result as dramatic deterioration in the material microstructure.

The electrostatic discharge can cause malfunction or total failure of an electronic component, Table 3.3. The failure can occur instantly or latently. The latter mode is typically due to increased parameter alteration caused by ESD breakdown in microelectronics. The cause of ESD-induced failure is typically related to improper handling or wrong material selections [24]. The ESD sensitive components or systems should contain a warning sticker, which is shown in Fig. 3.9.

A SAW filter component is an example of an ESD sensitive component. It might be failed with 50 V electrical discharges through it. To prevent ESD-induced failures, the product should be designed to tolerate them, i.e., by using alternate signal paths in case of ESD. Furthermore, the handling of the ESD sensitive devices should be properly organized. The handling of the components

Table 3.3 Failure mechanisms in IC microstructures related to electrostatic discharge [24]

Mechanism	Description
Bulk breakdown	Transistor parameter shifting due to breakdown in transistor microstructure. Breakdown path goes from Al-electrode through doped regions (P- or N-type) to Silicon substrate. High-current density causes alloying of semiconductor and precipitation of Al in the doped regions
Thermal secondary breakdown	Leakage current increase due to breakdown between PN-junction due to high voltage. This high speed current pulse generates very high temperature increase locally. Due to this, it damages the structure only in limited volume
Surface breakdown	Short circuiting or increase in leakage current due to breakdown between two adjacent metal conductors. The breakdown path progresses typically on the dielectric surface, hence the name
Dielectric breakdown	Short circuiting or increase in leakage current as a result of a high voltage breakdown through the dielectric material
Electromigration	Opens caused by a high electrical current density, which moves the atoms of the conductor in the direction of the electrons
Latch up	Transistor latches up due to ESD-pulse in transistor microstructure. This is due to undesirable biasing of PN-junctions

Fig. 3.9 A warning label of ESD sensitive component or product

includes human and machinery movements or operations. The component or the PWB can electrically be charged when it is moved or operated in SMT-line whenever the grounding breaks. This can be happening vice versa, so that machine is electrically charged and it could discharge through the PWB and components on it. ESD preventive actions in SMT production floor are presented in Table 3.4.

Table 3.4 Examples of ESD preventive actions in SMT manufacturing line

	Preventive action
1	Use only electrically grounded working places with static dissipative materials
2	Avoid or minimize the using or existence of charging materials at production floor
3	Make sure that the whole handling chain of the components or other subparts of the product, including human and machinery operations and movements is according to ESD protection policy
4	Operators must always use grounding wrists or heel straps when handling ESD sensitive products
5	Minimize handling
6	Give feedback to product design responsibilities to avoid the usage of ESD sensitive components or to take the ESD into account in the product design

3.5 Corrosion

It is said that corrosion can not be defeated but it can be stalled. Corrosion means a breaking down of essential properties of a material due to chemical reactions or mechanical erosion within it or its direct surroundings. There are multiple mechanisms of corrosion, e.g., uniform corrosion, pitting corrosion, crevice corrosion, and intergranular corrosion [7]. The galvanic corrosion is of particular concern in solder joints. Galvanic corrosion results to material loss in the anode metal when two metals with different electrochemical potential are galvanically connected. Electrolyte is essential in galvanic corrosion as its function is to carry the metal ions from the anode to the cathode metal. The chemical reactions at anode (M_1) and cathode (M_2) are as follows:

$$\text{Oxidation (anode): } M_1 \rightarrow M_1^{+} + 2e^{-}$$
$$\text{Reduction (cathode): } M_2^{+} + 2e^{-} \rightarrow M_2$$

Solder paste flux residues might accelerate galvanic corrosion when moisture is present. Corrosion will result to degrading of mechanical and electrical integrity of the signal paths.

Moisture (H_2O) plays a major role in corrosion mechanisms. Moisture is present in air and is then constantly present in electronics. The amount of moisture in air is expressed as relative humidity (RH). RH of an air–water mixture is defined as the ratio of the partial pressure of moisture vapor in the mixture to the saturated vapor pressure of water at the given temperature. The vapor pressure V_p at certain temperature T can be calculated from:

$$V_p = RH \times V_{S0} \times e^{\left[-\frac{\Delta H_V}{R}\left(\frac{1}{T}-\frac{1}{T_0}\right)\right]}, \tag{3.2}$$

where *RH is the relative humidity*, V_{S0} is the saturated water pressure at reference temperature T_0 and ΔH_V is the water heat of vaporization. As the RH increases the vapor pressure, the corrosion is accelerated.

At the IC microstructure level, there are several corrosion-induced failure mechanisms. Examples of these are (1) corrosion of aluminum metallization or wire, (2) corrosion of intermetallic compounds, (3) corrosion of gold wires, (4) corrosion of copper wires, and (5) corrosion from die bond material.

Moisture can also cause corrosion of aluminum metallization and generate hydrogen. The trapped moisture in cavities can be consumed by its chemical reaction with bare silicon that produces:

$$Si + 2\,H_2O \rightarrow SiO_2 + 2\,H_2. \quad (3.3)$$

Hydrogen can then aggravate the hot-carrier and radiation damage effects at the IC microstructure level. Thousand parts per million of moisture can already cause failures [21].

3.6 Plastic Package Popcorning

Some typical failure mechanisms related to plastic encapsulated component packages are presented in Fig. 3.10. In many failure mechanisms, the moisture plays a major role, as it is with corrosion failure mechanisms too. The ingressed moisture can cause swelling of plastic packages, which produces stresses that can cause shifts in electrical parameters, cracking, and delamination [10]. The moisture in plastic packages can also cause warpage and post-mold curing [25].

One of the most typical failure mechanisms is a delamination caused by so-called popcorning. When moisturized plastic packages are put to reflow oven, they might experience popcorning. The mechanism is described as following. Compo nent's plastic materials absorb moisture from the air. When the air exposure is long enough, the moist will penetrate into microscopic cavities in the component structure. As components are assembled to a PWB and reflowed the moisture starts to evaporate when the water boiling point is reached. The vapor pressure will get higher as the temperature rises toward the reflow peak temperature. At a certain point, the vapor pressure will exceed the strength of the laminated structure and delamination will occur, Fig. 3.11. The delamination will cause instant or latent failures to component.

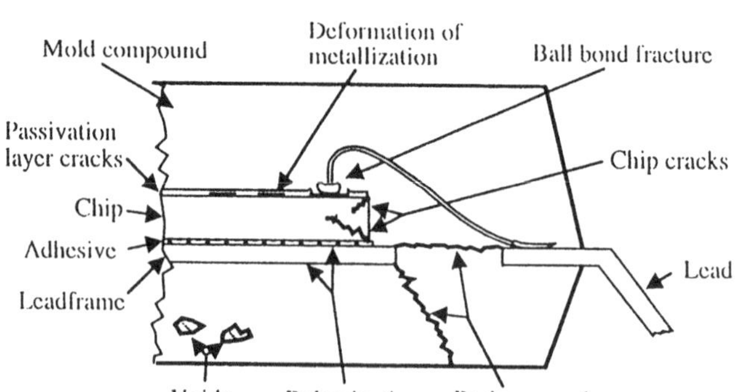

Fig. 3.10 A schematic example of failure mechanisms and sites in plastic component packages

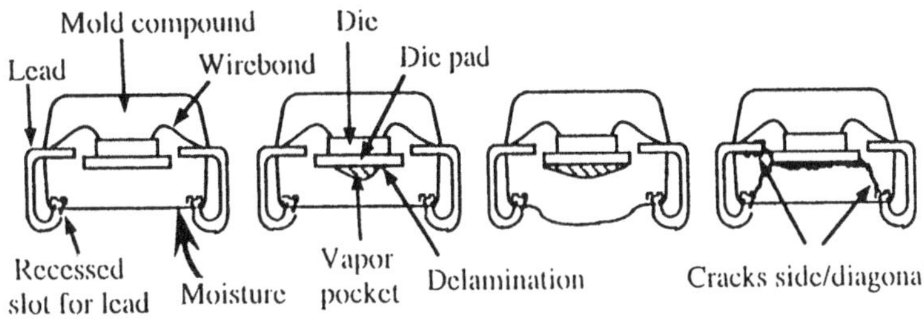

Fig. 3.11 Schematic presentation of plastic package popcorning

Exercises

3.1 What are the main categories and mechanisms of microsystem failures? Describe them briefly.
3.2 Why do electronic products fail? What are the typical failure causes?
3.3 One of the most common reasons for product returns is "no failures found," explain how can this be possible?
3.4 What kind of temperature cycling are mobile phones experiencing during a day? Where do these temperature changes come? Is the related stress causing fatigue, creep or brittle fracture of the phone's solder joints?
3.5 Describe the solder joint failure mechanism; brittle fracture. Give an example of the cause of a brittle fracture in solder joints.
3.6 Describe in detail the three stages of fatigue crack? Why and where fatigue must be taken into consideration when designing new electronics?
3.7 Describe in detail the three stages of creep? Why and where creep must be taken into consideration when designing new electronics?
3.8 How can electromigration be prevented in the ICs?
3.9 Describe Electrostatic charging and discharging.
3.10 What are the failure mechanisms related to ESD?
3.11 How can ESD failures be prevented in the SMT manufacturing line?
3.12 What are the failure mechanisms related to the presence of moisture in electronics?
3.13 How can moisture induced failure mechanisms be prevented or mitigated?
3.14 Think of any electronic device that you know for your personal experiences that have been failed during its operation.

(a) How did the failure occur? (Failure mode)
(b) Were the usage habitats according to product specifications (relates to overstress or slowly degrading failure mechanism)?
(c) Was the failure related to software, hardware, or system management?
(d) If you were the repairer of the product, how would you find the actual root cause of the product failure?
(e) How was the product repaired?
(f) If you were the product designer, how would you avoid the failure not to occur?

References

1. IPC-T-50G, "Terms and Definitions for Interconnecting and Packaging Electronic circuits". Standard by IPC, Lincolnwood, IL, 2003, 37.
2. RIAC, "Handbook of 217PlusTM Reliability Prediction Models", Department of Defense, USA, 2006, 146.
3. W. Brown, "Advanced Electronic Packaging with Emphasis on Multichip Modules", IEEE Press Series on Microelectronic Systems, 1999.
4. R. Wassink and M. Verguld, "Manufacturing Techniques for Surface Mounted Assemblies", Electrochemical Publications LTD, Bristol, 1995, 270.
5. W. Engelmaier, "Solder Joints in Electronics: Design for Reliability", April 2007, http://www.analysistech.com/downloads/SolderJointDesignForReliability.PDF.
6. J. Hwang, "Environment-Friendly Electronics: Lead-Free Technology", Electrochemical publications LTD, Bristol, 2001.
7. O. Vianco, "Corrosion Issues in Solder Joint Dedsign and Service", December 2007, http://www.osti.gov/energycitations/purl.cover.jsp?purl=/14961-8gkDYa/webviewable/.
8. T. Mattila, P. Marjamäki and J. Kivilahti, "Reliability of CSP Interconnections Under Mechanical Shock Loading Conditions", IEEE Transactions on Components and Packaging Technologies, 29(4), 2006, 787–795.
9. M. Klesnil and P. Lukas, "Fatigue of Metallic Materials", Materials Science Monographs, 7, Elsevier Scientific Publishing Company, Czechoslovakia, 1980.
10. F. Ellyin, "Fatigue Damage, Crack Growth and Life Prediction", p. 11 & 22. Chapman & Hall, London, 1997.
11. R. Wassink, "Soldering in Electronics", Electrochemical Publications LTD, Bristol, 1994.
12. J. Hwang, "Implementing Lead-Free Electronics", McGraw-Hill Professional Engineering, New York, 2005.
13. S. Stoeckl, A. Yeo, C. Lee and H. Pape, "Impact of Fatigue Modelling on 2nd Level Joint Reliability of BGA Packages with SnAgCu Solder Balls", IEEE, Electronics Packaging Technology Conference, 2005.
14. V. Lindroos, M. Sulonen and M. Veistinen, "Uudistettu Miekk-Ojan Metallioppi", Otawa, Helsinki, 1986, 738, 769.
15. A. Dasgupta and M. Pecht, "Material Failure Mechanisms and Damage Models", Proceedings of IEEE Transaction on Reliability, 40(5), 1991, 531–536.
16. S. Dunford, S. Canumalla and P. Viswanadham, "Intermetallic Morphology and Damage Evolution Under Thermomechanical Fatigue of Lead (Pb)-Free Solder Interconnections", Proceedings of IEEE Electronic Components and Technology Conference, 2004, 726–736.
17. S. Wiese, E. Meusel and K. J. Wolter, "Microstructural Dependence of Constitutive Properties of Eutectic SnAg and SnAgCu Solders", Proceedings of IEEE Electronic Components and Technology Conference, 2003, 202–205.
18. A. Schubert, R. Dudek, E. Auerswald, A. Gollhardt, B. Michel and H. Reichl, "Fatigue Life Models for SnAgCu and SnPb Solder Joints Evaluated by Experiments and Simulation", Proceedings of IEEE Electronic Components and Technology Conference, 2003, 603–610.
19. R. Dudek, H. Walter, R. Doering and B. Michel, "Thermal Fatigue Modelling for SnAgCu and SnPb Solder Joints", Proceedings of Thermal and Mechanical Simulation and Experiments in Micro-electronics and Micro-Systems, EuroSimE, 5, 2004, 557–564.
20. A. Prabhu, W. Schaefer and S. Patil, "High Reliability LTCC BGA for Telecom Applications", Proceedings of IEEE International Electronics Manufacturing Technology Symposium, 2000, 311–323.
21. A.G. Sabnis, VLSI Reliability, Ch. 6, Vol. 22, Academic Press, 1990.
22. P. Ho and T. Kwok, "Electromigration in Metals", Reports on Progress in Physics, 52, 1989, 301–348.
23. W. Boxleitner, "Electrostatic Discharge and Electronic Equipment", IEEE Press, New York, 1988.
24. T. Viheriäkoski, "ESD Staattinen sähkö elektroniikassa", EDITA, Helsinki, 2001.
25. T. Lin, Njoman, D. Crouthamel and K. Chua, "The Impact of Moisture in Mold Compound Preforms on the Warpage of PBGA Packages", IEEE, International Electronics Manufacturing Technology Symposium, 2003.

Chapter 4
Solder Joint Reliability

Abstract Solder joint reliability is the ability of solder joints to function under given conditions and to remain in conformance to both mechanical and electrical specifications for a specified period of time (without failing within the intended operating time).

In general, a particular failure mode is the result of certain failure mechanisms in which certain specific combinations of material properties and the surrounding environment act simultaneously. Many different factors have to be considered when assessing the reliability performance of a solder joint structure, such as stress distribution, strain amplitude, strain rate, the cyclic nature of the stress (mechanical, thermal, and thermomechanical), temperature, and many other environmental factors (corrosion, vibration, and so on). Apart from these, the metallurgical and physical behavior of the solder and the solder joint are also very important to take into account, since these also highly affect the reliability behavior of the solder joint.

The aim of this chapter is to increase the knowledge regarding reliability and failure of lead-free solder alloys/joints. This chapter gives an insight into how the microstructure of some lead-free solders is built its stability and some interfacial reactions. An introduction is also given to the failure mechanisms of solder joints, including fatigue failure, which is one of the most significant threats to the integrity of solder joints. Both the effect of second-level solder interconnection and some common standards used when testing solder joint reliability are also mentioned in this chapter.

4.1 Microstructure of Solder Joints

The microstructure of a solder joint is a combination of the grain structure and the phases present in the material, as well as defects, distribution, and morphology. The microstructure is dependent on solder alloy composition, substrate material, reflow time and temperature and solidification process, and thermal, mechanical, and chemical history of the solder material, from which cooling rate is one of the most critical factors. A faster cooling rate enhances the number of nuclei formed, giving smaller grains. With a slower cooling rate, the grains will be larger [1].

J. Liu et al., *Reliability of Microtechnology: Interconnects, Devices and Systems*,
DOI 10.1007/978-1-4419-5760-3_4,

During soldering, the formation of intermetallic compounds (IMCs) both inside the solder matrix and between solders and substrates is inevitable. The formation of intermetallic layers ensures good metallurgical bond and is therefore of utmost importance for the solder joint integrity and reliability. At low levels, intermetallics have a strengthening effect and produce distinct improvements in mechanical and thermal properties of solder alloys. However, at higher levels, they cause joint embrittlement [2]. A thick intermetallic layer, in surface-mounted solder joints, can be formed not only by a long reflow time and high reflow soldering temperatures, but also by aging, prolonged storage, and long-term operation of the assembly, even at room temperature. The IMC layer's thickness increases linearly with the square root of aging time, and the IMC/bulk solder interface gradually becomes flatter. A flat IMC/solder boundary is deleterious for the fatigue lifetime [3].

When soldering Sn-based solders on Cu, the most common IMC formed is Cu_6Sn_5, which forms when the molten solder wets the Cu [4]. It is common practice, however, to use electroless nickel and gold (ENIG) on top of the Cu surface as a barrier and oxidation protection layer between the solder and the Cu layer. The electroless immersion Au layer is very thin and is completely dissolved in the solder during soldering. Thus, wetting occurs toward the electroless nickel, which contains phosphors (P) and Ni_3Sn_4 IMCs will be formed between the Sn-rich solder and the Ni layer. Fractures have been observed at the Ni_3Sn_4/Ni–P interface, and possible reasons for these fractures are (1) the segregation of P at the interface, (2) contamination or oxidation during the Ni–Au plating or after plating via diffusion, and (3) brittle fracture of Ni–P and Ni_3Sn_4 [5].

The grain structure of the solder is also intrinsically unstable. The grains will grow in size over time as the grain structure reduces the internal energy of a fine-grained structure. A nonequilibrium microstructure will change to an energetically more favorable morphology over time [6]. This grain growth process is enhanced by elevated temperatures as well as strain energy input during cyclic loading. The grain growth process is thus an indication of damage accumulation [6]. After a certain time, microvoids can be found at the grain boundary intersections. These microvoids grow further into microcracks, which in turn grow into macrocracks leading to total fracture.

The properties of a material and the different mechanisms of solder joint failure are strongly influenced by the microstructure. Fatigue life, for example, can be drastically affected by the variation in microstructure. To improve the fatigue resistance of solder alloys, and from the cooling rate viewpoint, one should increase the cooling rate during solidification to create a more equiaxed microstructure [7].

4.1.1 Microstructure of Eutectic Sn–37Pb

The eutectic Sn–37Pb solder is a two-phase system, consisting of a mixture of soft lead-rich phase referred to as α-phase (solid solution of Sn in Pb) and tin-rich phase referred to as β-phase (solid solution of Pb in Sn). The eutectic reaction

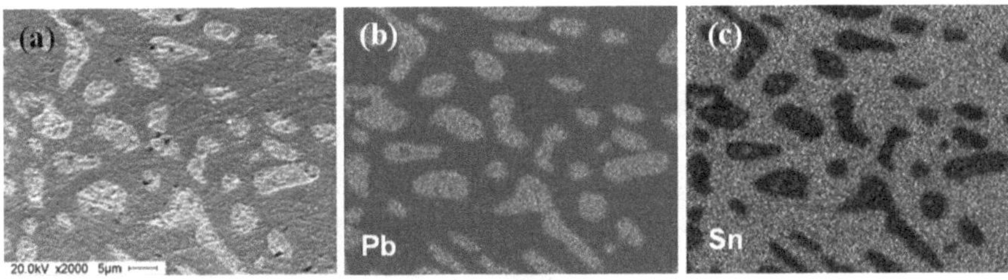

Fig. 4.1 (**a**) SEM/EDX micrograph of Sn–37Pb solder alloy microstructure; (**b**) Elemental distribution of Pb (*lighter areas*); and (**c**) Elemental distribution of Sn (*lighter areas*)

[L → (Pb α-phase) + (Sn β-phase)] takes place at 183°C (182.2°C) and is composed of 45.5wt% β-phase [(97.5−61.9)/(97.5−19.2) = 45.5] and 54.5wt% α-phase. At 183°C, the α-phase is composed of 19.2% Sn and the Sn-rich β-phase is composed of 97.5% Sn.

Figure 4.1a–c, shows the SEM/EDX elemental mapping of the eutectic Sn–37Pb, showing the distribution of Pb and Sn. The darker and lighter regions shown in Fig. 4.1a are the β-Sn-rich phase and the α-Pb-rich phase, respectively.

4.1.2 Microstructural Stability and Interfacial Interactions

Tin crystals have a body-centered tetragonal (bct) structure, which results in a thermal expansion difference along the principal axis of the crystals (α[100] = α[010] = 16.5 ppm/K and α[001] = 32.4 ppm/K, at 30°C) [8]; see Fig. 4.2a. Lead crystals, on the other hand, have a face-centered cubic (fcc) structure and behave isotropicaly with a CTE of 29 ppm/K, which is close to the maximum CTE of Sn crystals; see Fig. 4.2b. When the temperature is changed, the difference in CTE values between the Sn-rich and Pb-rich phases have to be accommodated internally by elastic and plastic strains. These plastic strains generate dislocations and consequently coarsening [9].

The microstructure of Sn–Pb alloys coarsens during its lifetime both under isothermal and thermomechanical loading conditions [10], and even at room temperature [11]. Current density also affects the microstructure of Sn–Pb solders, and substantial phase coarsening has been found to occur at 1×10^4 A/cm^2 and higher current densities [12, 13]. During coarsening the number of Pb-rich α-particles decreases and each particle becomes larger. This change in microstructure influences the mechanical properties of the solder, and consequently the mechanical response of the solder to loading dictates the lifetime and reliability limit of a circuit. The shear and fatigue strength of Sn–37Pb solder joints has been found to decrease with increased exposure to thermal cycling aging effects, which results in microstructural coarsening and IMC layers growth [14].

When soldered to Cu, only the Sn participates in the intermetallic formation with Cu, forming the typical Cu_6Sn_5 intermetallics. This results in the solder, at the

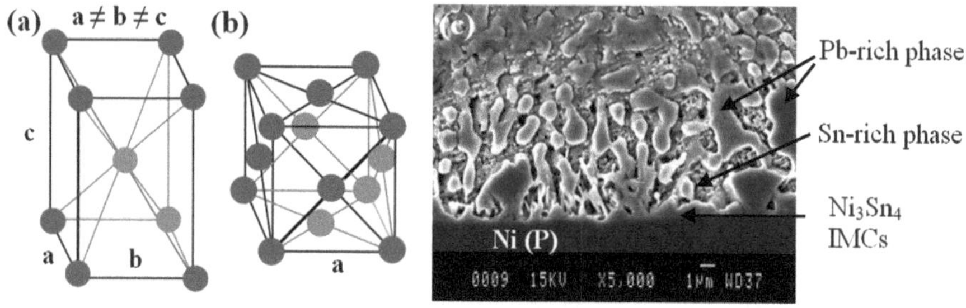

Fig. 4.2 (**a**) Schematic of bct crystal structure of Sn; (**b**) Schematic of bcc crystal structure of Pb; and (**c**) SEM micrograph of interface between eutectic Sn–37Pb and Cu/ENIG

solder/intermetallic interface, tending to be Sn-deficient (Pb-rich layer right above the Cu_6Sn_5 IMC layer), especially after long-time aging, since the Cu_6Sn_5 IMCs grow at the expense of the Cu from the UBM and the Sn from the solder. This behavior results in an interfacial failure mode for this type of solder alloy [15]. Sometimes, Cu_3Sn (ε-phase) is also found to form between the Cu and the Cu_6Sn_5 (η-phase) during thermal cycling [16]. When soldering on Ni (ENIG), the only intermetallics found are the Ni_3Sn_4 IMCs; see Fig. 4.2c. Sometimes, a P-rich layer can also be found between the Ni–P and IMC layer, since electroless Ni always contains a certain amount of phosphorous (P).

4.1.3 Microstructure of Eutectic Sn–3.5Ag

The microstructure of eutectic Sn–3.5Ag solder alloy consists of two phases, β-Sn and fine dispersed intermetallic Ag_3Sn particles within the β-Sn matrix; see Fig. 4.3. The eutectic reaction [L → Ag_3Sn + (Sn)] takes place at 220.3°C (~221°C) and the liquid has a mass fraction of 3.73% Ag and 96.27% Sn, the Sn-phase is 99.93% pure Sn (0.07% Ag), and the Ag_3Sn phase is composed of 73.17% Ag and 26.83% Sn. Furthermore, there is no solid solubility of Ag in Sn. In addition to the fine Ag_3Sn intermetallics, large needles of Ag_3Sn can also be present. Too large Ag_3Sn particles are not desirable and to avoid this, microstructural refinement achieved through the addition of other small particles (rare earth elements) has been done.

4.1.4 Microstructural Evolution and Interfacial Interactions

When soldering Sn–3.5Ag to Cu/ENIG finish, the only IMCs were found at the interface as the binary Ni_3Sn_4; see Fig. 4.4a. Although coarsening of the Ag_3Sn particles was observed as a function of thermal cycling, no significant increase in the thickness of the Ni_3Sn_4 IMC layer was observed; see Fig. 4.4b.

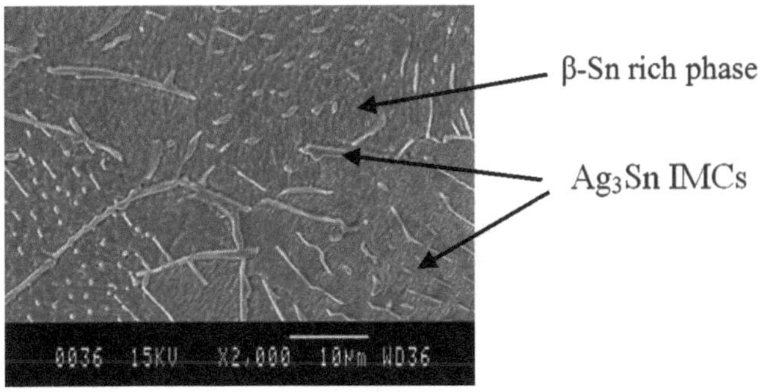

Fig. 4.3 SEM micrograph of eutectic Sn–3.5Ag solder alloy (as reflowed)

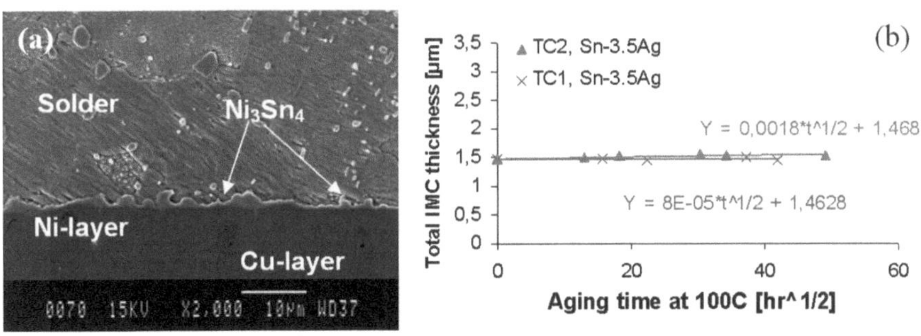

Fig. 4.4 (**a**) SEM micrograph of interface between Sn–3.5Ag solder and Cu/ENIG finish, thermal cycled for 7,000 cycles at TC2; (**b**) Total IMC layer thickness versus aging time (at 100°C) for two thermal cycling profiles, TC1 and TC2

4.1.5 Microstructure of Sn–Ag–Cu Alloys

The Sn–Ag–Cu (SAC) alloys have a structure of tin-rich dendrites with Ag_3Sn and Cu_6Sn_5 IMCs dispersed throughout; see Fig. 4.5a. Figure 4.5b shows that these IMCs are normally found at the Sn-grain boundaries. The Ag_3Sn IMCs can also be formed as large plates, normally attached to the interfacial intermetallics or voids. The eutectic reaction [L $\rightarrow$ Ag_3Sn + Cu_6Sn_5 + (Sn)] of the SAC system is said to take place at 216°C (215.9°C) with a composition of Sn–3.7Ag–0.9Cu, at which temperature the Ag_3Sn-phase is composed of 73.17% Ag and 26.83% Sn, the Cu_6Sn_5-phase is composed of 39.07% Cu and 60.93% Sn, and the Sn-phase is composed of 99.93% Sn and 0.07% Ag.

Figure 4.5c–e shows the SEM/EDX elemental mapping analysis of the IMC particles, including the elements Ag, Cu, and Sn found in the Sn–4.0Ag–0.5Cu solder microstructure. By means of EDX, the larger particles were identified to be an Sn–Ag phase with an average compositional value of Sn:Ag = 29.3:70.7 and these IMCs could be denoted as Ag_3Sn. Many other researchers have shown that Ag_3Sn particles are evenly distributed among the eutectic colonies or along the

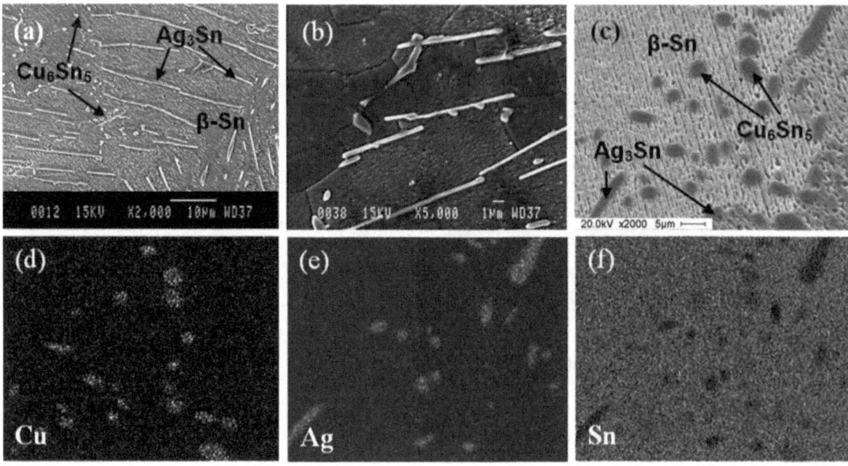

Fig. 4.5 (**a**) SEM micrograph of Sn–4.0Ag–0.5Cu solder alloy; (**b**) IMCs at the grain boundaries; (**c**) SEM micrograph for EDX elemental mapping of small Ag_3Sn and Cu_6Sn_5 intermetallics found in the SAC alloy; (**d**) Elemental distribution of Cu; (**e**) Elemental distribution of Ag; and (**f**) Elemental distribution of Sn

Sn-rich phase boundaries in the Pb-free systems [17, 18]. The morphology of the Ag_3Sn intermetallics depends on the cooling rate; slow cooling rate results in the formation of larger needle-like Ag_3Sn particles, while fast cooling rate results in finer Ag_3Sn particles. Small Ag_3Sn particles have been identified to reinforce the solder matrix [18] and to improve the mechanical properties of the solder alloy. The presence of such particles reduces, however, the solder joint's ductility by inducing brittle fracture mode, and influencing crack initiation, and accelerating the fatigue crack growth kinetics due to decohesion of large Ag_3Sn particles, especially when they have a branch-like morphology that deteriorates the homogeneity of the mechanical properties [19, 20]. The finer particles dispersed in the SAC alloy were identified as being a Cu–Sn phase composed of 53.6 at% Cu and 46.4 at% Sn indicating the Cu_6Sn_5 phase.

4.1.6 Microstructural Evolution and Interfacial Interactions

For the thermal cycling conditions used in the present work, and as shown in Fig. 4.6a, b, the microstructure of SAC alloy changes as the IMCs become coarser as a function of thermal cycles. Coarsening has also been observed by other researchers, especially as a function of isothermal aging. At a higher aging temperature of 180°C, coarsening of the Ag_3Sn particles was also observed. Room temperature aging tests performed on Sn–3.9Ag–0.6Cu have also showed a continuous material softening which was correlated to the growth of relatively large tin-rich crystals [21].

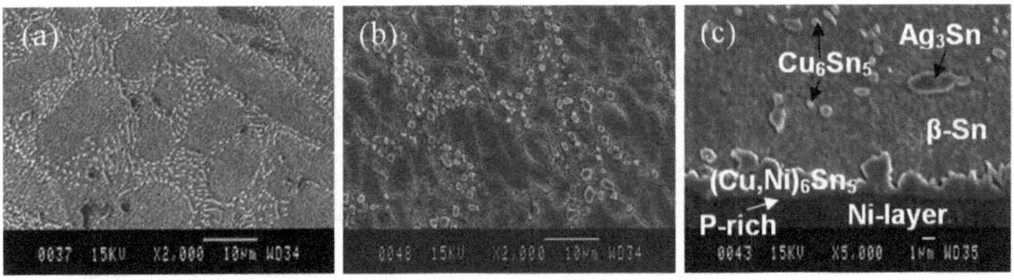

Fig. 4.6 SEM micrograph of a reflow soldered 0805 solder joints using SAC, (**a**) as reflowed, before thermal cycling; (**b**) tested for 5,500 cycles at TC1; (**c**) interface between the SAC alloy and ENIG

Fig. 4.7 Relationship between the average thickness of the IMC layer and the square root of aging time for both surface-mounted and wave-soldered 0805 components, tested at both TC1 and TC2

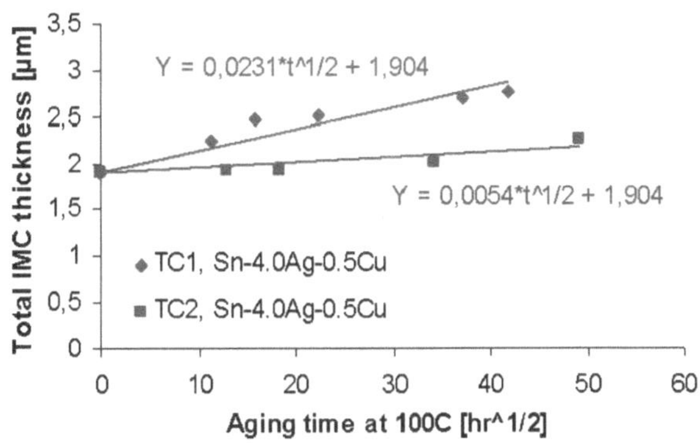

One reason for room temperature microstructural changes of this alloy is that at 25°C, the diffusion coefficient of Cu along the "a" and "c" axes of Sn is approximately 0.5×10^{-8} and 2×10^{-6} cm^2/s, respectively, indicating a sufficiently high mobility of the Cu atoms in the Sn to enable the growth of the Cu–Sn intermetallic phase [22].

When soldering SAC on ENIG, the intermetallic layers between the solder and the ENIG are only composed of $(Cu, Ni)_6Sn_5$; see Fig. 4.6c. The thickness of this layer was measured as a function of the square root of aging time both for TC1 (ranging between −55 and 100°C, with a ramp rate of 10°C/min and a dwell of 15 min at both temperature extremes, resulting in a period time of 61 min) and TC2 (ranging between 0 and 100°C, with a ramp rate of 10°C/min and a dwell of 10 min at both temperature extremes, resulting in a period time of 40 min), and the results are shown in Fig. 4.7. A measurable increase in IMC layer thickness, as a function of the square root of aging time, was observed for TC1.

4.1.7 Microstructure of Sn–3.5Ag–3Bi

Figure 4.8a shows an SEM micrograph of as-reflowed Sn–3.5Ag–3Bi solder alloy. The microstructure of this alloy is composed of a β-Sn-rich matrix with some

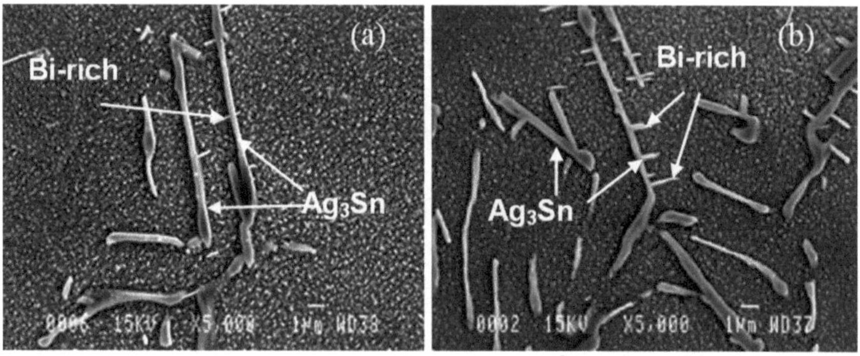

Fig. 4.8 SEM micrograph of Sb–3.5Ag–3Bi alloy, (**a**) as reflowed and (**b**) after room temperature LCF testing

dispersed Ag_3Sn IMCs and some Bi-rich particles. Bismuth does not build any IMCs with either Sn or Ag, and that is why Bi is found as single fine particles dispersed in the β-Sn-rich matrix. Solder alloys containing up 2% Bi show smaller β-Sn dendrites and more irregular Ag_3Sn precipitates. Adding Bi over the solid solubility limit to the tin-rich alloy results in the crystallization of fine Bi and irregular-shaped Ag_3Sn IMCs around β-Sn globules containing Ag_3Sn IMCs [23].

4.1.7.1 Microstructural Evolution and Interfacial Interactions

As shown in Fig. 4.8b, Bi crystallization was also found after the LCF tests at room temperature performed in this work. In this case, some Bi-rich particles crystallized around the Ag_3Sn IMC particles. Since no intermetallics are formed between Bi and Sn and Ag, the only IMCs expected to be found at the interface between this alloy and ENIG are Cu_6Sn_5 with some diluted Ni inside, resulting in $(Cu, Ni)_6Sn_5$.

Experiments performed on solder alloys containing Bi show that the higher the Bi content, the lower the IMC layer growth as a function of aging time [24]. The Bi-rich particles have shown some coarsening as a function of isothermal aging [25].

4.1.8 Microstructure of Sn–0.7Cu–0.4Co

The microstructural analysis of a Sn–0.7Cu–0.4Co solder alloy reveals that the only IMCs found between the Co and Cu, Sn or Ag are rod-like $CoSn_2$. The reason for this is that Co has little solubility in the β-Sn matrix, Ag and Cu and, therefore Co does not form any IMCs with either Ag or Cu [26, 27]. Figure 4.9a–d shows the

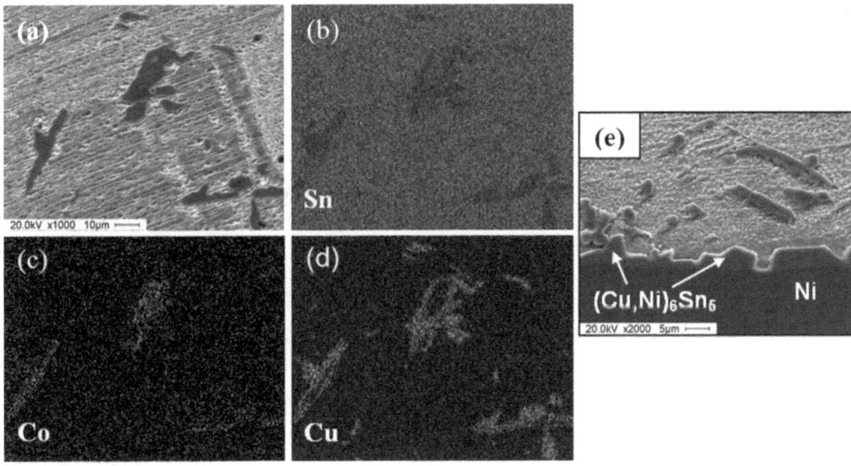

Fig. 4.9 EDX elemental mapping of Sn-0.7Co-0.4Cu alloy, (**a**) SEM micrograph of analyzed area, (**b**) distribution of Sn, and (**c**) distribution of Co, and (**d**) distribution of Cu, and (**e**) Interface between Sn-0.7Cu-0.4Co solder and ENIG board finish

SEM/EDX mapping analysis of the Sn–0.4Co–0.7Cu solder alloy, showing the IMCs, including Cu and Co elements. According to the EDX analysis, the atomic composition of the Co-rich particles is Sn:Co:Cu, 70:26:4 at% corresponding to the $CoSn_2$, with the substitution of Cu (solid solution) into the $CoSn_2$ phase, resulting in $(Co,Cu)Sn_2$. The Cu-rich IMCs show an atomic ratio between Sn:Cu:Co of 59.5:34.5:6 and can therefore be identified as the $(Cu,Co)_6Sn_5$ IMC phase (Cu_6Sn_5 phase with the substitution of Co). The microstructure of the Sn–0.7Cu–0.4Co alloy is therefore composed of a β-Sn-rich matrix and dispersed $CoSn_2$ and Cu_6S_{n5} IMCs.

4.1.8.1 Microstructural Evolution and Interfacial Interactions

Aging tests at 150°C for 24 h were performed to investigate the stability of the Sn–0.7Cu–0.4Co alloy. The results of these tests show no apparent changes in the microstructure between the as-solidified and the aged material state. Both the Cu_6Sn_5 and the $CoSn_2$ IMCs seem to be of the same size and shape and have the same distribution in the Sn-rich matrix. This alloy, therefore, presents a relatively stable microstructure (under the present aging conditions). It is known, however, that the solubility of Cu in Sn is quite high, and therefore it would be expected that at least some changes in the Cu_6Sn_5 particles would be observed. The introduction of Co, however, which is also present in the Cu_6Sn_5 IMCs, might hinder the movement of the Cu atoms since the vacancies are already taken by the Co atoms.

When soldering Sn–0.7Cu–0.4Co on Cu/ENIG, the only IMCs found were the binary phase Cu_6Sn_5 with some Ni diluted inside resulting in the ternary $(Cu, Ni)_6Sn_5$ interfacial compounds [27]; see Fig. 4.9e.

4.2 Mechanical Reliability of Solder Joints

In the assessment of reliability performance of a solder joint structure, the stress distribution, the strain amplitude, the strain rate, the cyclic nature of the stress (either mechanical or thermal), the temperature, and other environmental factors all have to be considered. Basic processes or factors that are believed to be probable reasons for solder failure while in service are inferior or inadequate mechanical strength, creep, mechanical and/or thermal fatigue, thermal expansion anisotropy, corrosion-enhanced fatigue, IMC formation, detrimental microstructure development, voids, electromigration, and leaching [7].

A particular failure mode is the result of a certain failure mechanism in which certain specific combinations of material properties and the surrounding environment act simultaneously. There are three major mechanisms of solder joint failure, namely, tensile rupture (fracture due to mechanical overloading), creep failure (damage caused by a long-lasting permanent load or stress), and fatigue (damage caused by cyclical loads or stresses). These three mechanisms often interplay simultaneously with each other. Most of the physical failure modes (fatigue, delamination, creep, etc.) are generally a result of thermomechanical stresses. There are, however, other factors that also affect the failure behavior of solder joints. Electrical and chemical actions can also be responsible for many thermo-mechanical failures in modern electronic packages. Electromigration-induced voiding is for example primarily due to high-electrical current density. Corrosion is another factor that accelerates fatigue and delamination failure.

A solder joint alone cannot be considered as reliable or unreliable according to the definition of reliability above; it becomes so only in the context of a micro-system where the components are connected via the solder joints to the PCB. The properties of the component, substrate, and solder joint, together with service conditions, design life span, and the acceptable failure probability for the whole assembly determine the reliability of the surface mount solder attachment.

Surface-mounted solder joints provide electrical, mechanical, and thermal functions and should, for that purpose, be ductile enough to deform and withstand different levels of stresses and strains. Solder joints are far from being a homogeneous structure and its microstructure is normally very complex consisting of different layers such as the base metal at the PCB, followed by one or more IMC layers, then a layer from which the solder constituent forming the PCB-side IMCs has been dissipated, followed by the solder grain structure. The same type of structure is found on the component side. The typical solder joint structure of a BGA solder ball is shown in Fig. 4.10.

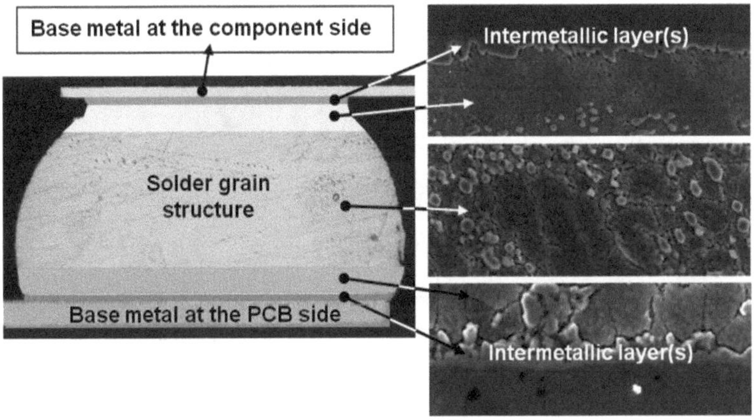

Fig. 4.10 BGA solder joint structure depicting the different layers

4.2.1 Fatigue Failure

Fatigue failure is one the most significant threats to the integrity of solder joints. Fatigue is defined as a measure of resistance to cyclic loading, which can either be mechanical or thermal (temperature fluctuations). Fatigue, which occurs under alternating stresses, causes failure when the maximum stress is of sufficiently high value, or there is enough variation of the applied stress, or even if there is a sufficiently large number of stress cycles. Fatigue leads sooner or later to crack initiation and propagation and finally failure. Fatigue slowly degrades the integrity of a soldered interconnection until it becomes an electrically open path. On the atomic scale, when a metal is subjected to cycling loading, atomic motion and rearrangement by plastic flow will occur, resulting in hardening or softening, depending on the material [7]. Since fatigue failure may occur at much lower cyclic stresses than would be required for a single (static) load application, fatigue failures are immature and unexpected, and therefore a very important issue to be analyzed in the context of solder joint reliability.

Microsystem packages comprise dissimilar materials that expand at different rates on heating as a result of different coefficients of thermal expansion (CTE). The CTE mismatch between the different materials, in combination with temperature cycling caused by either on or off cycles, daily temperature variations, and seasonal changes impose a significant cyclical inelastic (plastic) strain into the solder joints that ultimately will lead to fatigue failure of the solder joints. Even if the CTE were exactly equal in all components, the difference in thermal expansion would persist due to the fluctuations in the temperature gradients across the device under conditions of power cycling [8].

Figure 4.11a–c shows how temperature fluctuations produce significant strain levels as a consequence of CTE mismatch. Following the temperature variations, the strains are cyclic in nature with a frequency determined by the operation profile,

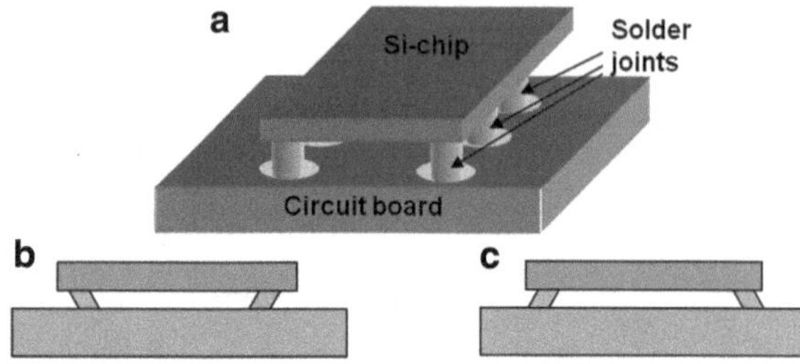

Fig. 4.11 Schematic picture of the development of strains in a microsystem device; (**a**) microsystem at zero strain; (**b**) system with negative temperature excursion; (**c**) system with positive temperature excursion

which additionally involves the incorporation of hold and dwell periods in the cycle. Since the solder is usually the softest material in the package, the developed strain is concentrated at the solder joints and may ultimately cause failure by fatigue in shear.

The free expansion of two substrates A and B, with CTE values α_A and α_B, is due to a temperature increase from T_0 to T:

$$\begin{cases} d_{th(A)} = (T - T_0)\alpha_A l \\ d_{th(B)} = (T - T_0)\alpha_B l \end{cases}, \tag{4.1}$$

where l is the total length of the device. When these two substrates are joined by solder, their free thermal expansions are prevented and there is an interaction between the substrates and the solder. The thermal expansion of the solder joint is then the difference in thermal expansion between substrate A and B, as:

$$d_{th} = (T - T_0)\alpha_A l - (T - T_0)\alpha_B l. \tag{4.2}$$

If the CTE $\alpha_A < \alpha_B$ and $T > T_0$, substrate A experiences a tensile displacement and substrate B a compressive displacement.

Solder joint fatigue resistance can be increased by using ductile solder alloys, which present a capacity to deform before they fracture. Creep resistance can be increased by using solder materials with higher melting temperatures.

4.3 General Solder Joint Failure Mechanism

In the literature, there are numerous studies concerning the solder joint failure mechanisms. The solder joint failure (solder crack) is fundamentally a low-cycle fatigue mechanism, which follows the well-known Coffin–Manson relation. Low cycle fatigue is considered to be related to "plastic-strain fatigue." However, three sources for solder cracking can be distinguished, which are overload, long lasting permanent load and cyclic load. The causes for the latter two are creep and fatigue, respectively. Some estimates about the usage conditions and stress levels experienced by the components and their solder joints have been published. In the references, the stress conditions are divided into different device type and environment categories, which also describe general level requirements for the devices. The approach is in the product category level and the stress variation of individual products is not taken into account.

To predict the solder joint failure occurrence under field conditions, the stress levels in the field must be known. Furthermore, an interpretation method to utilize the accelerated stress test data for failure prediction is needed. There have been some reported prediction methods for estimating the second level interconnection failures. These models are based on the empirical data and are relatively accurate. What is common to these prediction methods is that they have unfortunately not been taken into wider use in the product level reliability analysis. The aforementioned methods have a great deal of data and analysis behind them, but their usage has not found its way to the real Mean Time To Failures (MTTF) calculations as such, which means, in practice, that they are not a solid part of product development activities.

The fundamental idea behind accelerated stress testing is first to identify the stress conditions in the field environment. In the test phase, the stress levels are then increased to accelerate the failure occurrence. Correctly chosen test parameters ensure the shortest test time and correct analysis. This approach will result in improved reliability in product design. However, randomly selected test set-up usually leads only to wasting the research resources.

The failure definition is one of the key tasks in the accelerated stress tests. For example, the solder joint failure can be defined as an irreversible change in electrical resistance. For example, IPC-9701(2002) specifies that the joint has failed when 20% resistance increase has been detected with data logger within five consecutive scans.

Usually when the first anomalies in the continuous monitoring data are detected, no macroscopic deformations can be observed. In spite of that, when looking deeper into the solder joint, the fundamental analysis would point the failure root cause to the deformations in the grain or grain boundary level, meaning that the solder material has aged. The next step, further inside the solder material, would go to the atomic level research. One could assume that atomic level research would give more answers about the aging characteristics of the solder. However, as is fundamentally known, the atoms do not age according to traditional physics. Thus it could be stated that the smallest aging elements in the solder joint system are the crystal lattices and their defects.

Another conclusion is that a failure can be indirectly detected by electrical measurements without going into the analysis, for instance, in the crystal level of material. By widening this idea, the properties of material can be divided into two categories, so-called intrinsic material properties and extrinsic material properties. The extrinsic properties are such macrolevel properties which can be directly or indirectly measured in the macrolevel. The intrinsic material properties are the foundation of the extrinsic properties and cannot usually be directly measured.

Table 4.1 Material properties and phenomena of a solder joint system (Hwang 2001: 65–69)

Property type	Examples of the properties
Physical	Phase-transition temperature
	Electrical conductivity
	Thermal conductivity
	Coefficient of thermal expansion
	Surface tension
Mechanical	Stress–strain behavior
	Creep resistance
	Fatigue resistance
Metallurgical phenomena	Plastic deformation
	Strain-hardening
	Recovery process
	Recrystallization
	Solution hardening
	The recovery process

Some of the most important properties of a solder joint system, including some important metallurgical phenomena are presented in Table 4.1. From a solder joint mechanical stability point of view, the stress–strain behavior, the creep resistance, and the fatigue resistance are the most important properties.

The aging mechanisms of the solder joints depend on the stress conditions and the thermomechanical properties of the soldering system. The solder joint material does have multiple properties that affect the failure occurrence. The aging of the solder takes place in the intrinsic material property level and it might not be detectable in macroscopic level before final material property degradation. The material aging models have been developed to estimate the failure occurrence in different stress conditions. To model the aging of solder, constitutive models of solder material have been developed.

Within the solder aging models, the solder deformation rate, or the solder strain rate, is divided into substrain rates of elastic, viscoelastic, creep, and plastic deformations. For low cyclic loadings of solder, with low temperature ramp rates, the creep is the most important part of strain, dominated by secondary, or steady-state, creep. The strain rates in the secondary creep are close to constant, hence the name steady-state creep. Creep is the result of an applied static stress, in which the material relieves the stresses with plastic deformation.

The plastic deformation is caused, among others, by the secondary creep, which is controlled, e.g., by the movement of dislocations in the slip planes. Creep in solder is due to dislocation climb mechanism or due to grain boundary sliding and by intergranular or transgranular void migration (grain boundary diffusion). Dunford et al. [28] published that most relevant creep deformation mechanisms with lead-free solders are dislocation creep, diffusion creep, and grain boundary sliding. In recent years, many lead-free creep constitutive models have been developed, which emphasize the importance of understanding the creep mechanisms behind the solder joint failures. When solder joints are subjected to cyclic stress environments, e.g. thermal cycling, the solder joints can fracture at stress levels below the

Table 4.2 Mechanisms of crack nucleation

Mechanism
Coarse slip on alternating parallel slip planes
Local brittle fracture
Condensation
Loss of coherency across a slip plane due to accumulation of defects
Nucleation of cracks in grain boundaries

yield strength. This solder joint fatigue is based on a plastic deformation in the microscopic level, whereas it is not observed in the macroscopic level. The fatigue failure can be divided into three phases (1) crack nucleation, (2) crack propagation, and (3) final fracture. The crack nucleation is preceded by microstructural changes, e.g. local grain growth. Examples of crack nucleation mechanisms are presented in Table 4.2. The crack propagation starts with crystallographic propagation, which is followed by noncrystallographic propagation. The latter has a faster propagation rate and only one crack is usually propagating.

One mechanism behind the solder fatigue is the recrystallization followed by the grain growth, where the contaminants and microvoids coalesce in the grain boundaries and weaken the mechanical properties of the solder joint. The fracture can nucleate and propagate from the defects in the grain boundaries and eventually end as a full fracture. Coffin and Manson explained the fatigue crack growth in the terms of plastic strain. This generally used relation for low cycle fatigue is expressed by:

$$N(\Delta\varepsilon_p)^n = C_{pf}, \tag{4.3}$$

where N is the number of cycles to failure, n is an empirical constant, $\Delta\varepsilon_p$ the plastic strain range during one cycle, and C_{pf} is a proportionality factor (Norris and Landzberg [29]). By using (4.3) comparison between different strain ranges can be made. The Coffin–Manson relation has been accepted as a basis for many solder fatigue models. The Coffin–Manson relation is also the basis of so called Norris–L andzberg relation. The Norris–Landzberg relation was recently reviewed by Salmela and by Pan et al. [30, 31] to update it to correspond with the SnAgCu solder. Examples of stress conditions and failure mechanisms experienced by solder joints are shown in Table 4.3. As stated, the creep and fatigue are the failure mechanisms for the solder joint system, which are typical examples of wear-out failures of solder joints. On the other hand, brittle fractures are typical in the shock environments, i.e. high acceleration caused by drop. To accelerate the primary failure mechanisms of solder joints, thermal fatigue and creep, thermal cycling, and vibration-based tests are used.

The thermal cycling, as a reliability test method, has been in wide use for decades. The vibration tests are usually done in the product verification phases, at least for the infrastructure electronic products. The shock and drop tests are commonly used for the devices, which might experience such stress conditions during their life-time. For example, cellular phones are tested against the drop shock loadings. The products are also tested against the corrosive field environments, but the solder

Table 4.3 An example of stress conditions experienced by solder joints and the related failure mechanisms

Failure mechanism	Stress environment	Description
Fatigue[a]	Cyclic stress, thermal[a] or vibration[b]	Fatigue failure initiates with microcrack and propagates to solder crack in cyclic stresses[b]
Creep[a]	Long-lasting permanent load[a]	Global plastic deformation of solder under static mechanical stress and temperature[c]
Corrosion[b]	Galvanic pair[b]	Metal with different electrochemical potential is in contact to cause material loss in the anode metal[d]
Brittle fracture[e]	Drop or shock[e]	Fracture occurs in the brittle intermetallics layer of solder joint[e]

[a]Wassink and Verguld [32]
[b]Engelmaier [6]
[c]Hwang [33]
[d]Vianco [34]
[e]Mattila et al. [35] and Prabhu et al. [36]

material itself is not typically the first to fail. So in the bottom line, the thermal cycling test performance as a measure of reliability has been accepted as a solid method to test the aging mechanisms of the solder joints.

4.3.1 Effect of Second Level Solder Interconnection Failure

Interconnection failures are observed at first as degradation in the product performance and will inevitable lead, after all, to product failure. This will happen if the redundancy was used in the design or not. Figure 4.12 shows measured electrical resistance over four solder joints of one component in the accelerated stress test. One can note that after stable 450 h in the test, there is an approximately 100h region, when the aging of the solder joint can be detected by the resistance measurements. This is due to solder joint crack which has started to propagate, Fig. 4.13. In this region, the product performance starts to degrade and might cause short periods of product malfunction. If the product is sent to repair, a failure may not be detected at all (No Faults Found). Moreover, with the radio frequency/ microwave applications, the degradation of the signal integrity can be detected with the S-parameter measurements as an increase in return loss in the crack propagation phase. When the solder joint crack propagates further to full rupture, there will be two surfaces very close to each other in both sides of the solder joint rupture, Fig. 4.14. During the thermal fluctuations, the contact area of these two surfaces is continuously changing. As a result of this, the resistance over the joint will be unstable. This behavior is also assisted with the surface contamination. After all, the product will fail as a result of the solder joint aging. To prevent the failures, the solder joints should stay in the stable region with a certain safety limit.

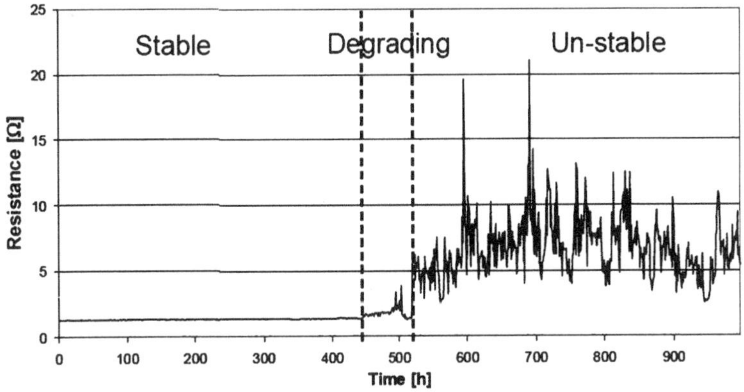

Fig. 4.12 Measured electrical resistance over four solder joints of a ceramic leadless component as a function of time in the accelerated stress test

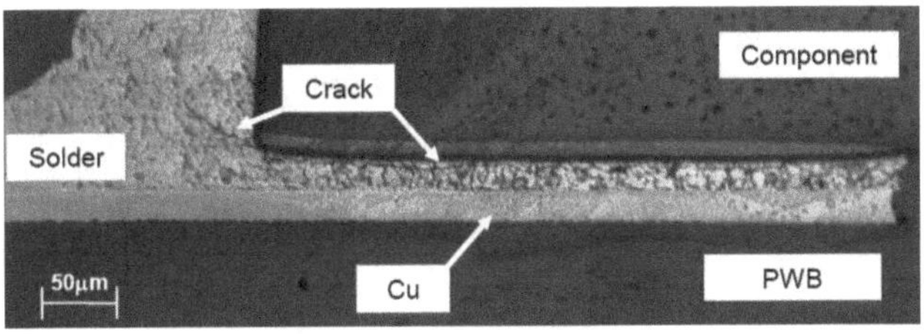

Fig. 4.13 Microsection of SnPb solder joint of a ceramic leadless component after accelerated stress test. Crack paths are already starting to propagate. The electrical and mechanical properties of the solder joint have degraded. Such solder joints might be seen as degraded product performance

Fig. 4.14 Microsection of SnPb solder joint of a ceramic leadless component after accelerated stress test. The crack path has fully developed. The electrical and mechanical properties of the solder joint are totally gone, or at least they are very unstable. Note that the electrical and mechanical connect might still exist in some part of the solder joint; figure shows only two-dimensional microsection of the solder joint

RIAC [37] has collected failure causes of electronic systems. In the study, there is no particular category for the interconnection failures. This is mostly due to the fact that the failures are not investigated, and only the initial symptom of the failure is recorded. The interconnection failures can usually be found under the manufacturing, wearout or design categories, but they can be a part of any other category [37]. This study is in line with other similar studies concerning the missing analysis of solder joint failures. As a conclusion, there is no accurate record of the solder joint failures. This results in the predictions relying predominantly on simulation methods.

4.3.2 *Standards Related to Solder Joint Reliability Testing*

As the thermal cycling has been accepted as a common component and board-level reliability test method, there has been a need for the standardization of the test methods. The Association Connecting Electronics Industries (denoted as IPC for its former name) has a wide selection of standards for the reliability test methods and requirements. Table 4.4 shows the IPC specifications for the tests of board-level interconnections. The IPC-9701 is a widely used reference in the electronics industry, even though it does not explicitly define the requirements for the thermal cycling test characteristics. Table 4.5 shows temperature cycling condition options specified by IPC 9701 (TC1–TC5). The test setup and result requirements of IPC-9701 are made for qualification purposes and the data are not currently used for field failure predictions.

Table 4.4 Examples of specifications of component interconnection tests

Standard	Description of the standard
ICP-9701	Solder joint reliability-performance test methods and qualification requirements for surface mount solder attachment
ICP-9702	Monotonic bend test-monotonic bend characterization of board-level interconnects. For resistance to strain
ICP-9703	Mechanical shock test methods and qualification requirements for surface mount solder attachments
ICP-D-279	Design guidelines for reliable surface mount technology printed board assemblies

Table 4.5 Temperature cycling conditions per IPC-9701(2002)

Minimum temperature [°C]	Maximum temperature [°C]	Test condition
0	100	TC1
−25	100	TC2
−40	125	TC3
−55	125	TC4
−55	100	TC5

Exercises

4.1 Why is there so much concern about lead?
4.2 Why is lead (Pb) being phased out? (Motivation)
4.3 Is there legislation that bans Pb use in electronics?
4.4 What are the advantages of using Pb-free packages? What effort in eliminating lead-containing solders is coming from Europe?
4.5 What is the definition of lead-free?
4.6 What problems will meet when converting to lead-free?
4.7 Which changes will be necessary in customer processes when using lead-free components?
4.8 In the reflow process, is there any necessary to modify printing parameters or stencil design for lead-free?
4.9 Do higher soldering temperatures have any negative impact on the moisture sensitivity level (MSL)?
4.10 Describe the concept of thermomechanical design of electronic packages as an up-front design activity for screening out and minimizing process and reliability-related failures.
4.11 Electronic packaging material properties such as the elastic modulus, $E(T)$, yield stress, σ_y (T), and coefficient of thermal expansion, CTE(T) are dependent on temperature. How these properties affect the thermomechanical reliability performance of solder joints in electronic assemblies subjected to thermal cycling loading?

References

1. M. Abtew, "Lead free solders for surface mount technology applications (Part 1)," Chip Scale Review, 1998, http://www.chipscalereview.com/9809/m.abtew1.htm.
2. P. G. Harris, K. S. Chaggar, "The role of intermetallic compounds in lead-free soldering," Soldering & Surface Mount Technology, 10(3), 1998, 38–52.
3. Y. C. Chan, P. L. Tu, A. C. K So, J. K. L. Lai, "Effect of intermetallic compounds on the shear fatigue of Cu/63Sn–Pb solder joints," IEEE Transactions on Components, Hybrids and Manufacturing Technology-Part B, 20(1), 1997, 87–93.
4. Z. Guo, P. Hacke, A. F. Sprecher, H. Conrad, "Effect of composition on the low-cycle fatigue of Pb alloy solder joints," 40th Electronic Components and Technology Conference, Las Vegas, NV, 1, 1990, 496–504.
5. H. D. Blair, P. Tsung-Yu, J. M. Nicholson, "Intermetallic compound growth on NI, Au/Ni, and Pd/Ni susbstrates with Sn/Pb, Sn/Ag, and Sn solders," 48th Electronic Components and Technology Conference, Seattle, WA, 1998, 259–267.
6. W. Engelmaier, "Solder Joints in Electronics: Design for Reliability", in Design and Reliability of Solders and Solder Interconnections, Edited by R.K. Mahidhara et al. The Minerals, Metals & Materials Society, Gaithersburg, MD, 1997, 9–13.
7. J. W. Morris, H. J. Reynolds, "The Influence of Microstructure on the Failure of Eutectic Solders" in Design and Reliability of Solders and Solder Interconnections, Edited by

R. K. Mahidhara, S.M.L. Sastry, P.K. Liaw, K.L. Murty, D.R. Frear, and W.L. Winterbottom, The Minerals, Metals & Materials Society, Gaithersburg, MD, 1997, 49–58.
8. M. A. Matin, W. P. Vellinga, M. G. D. Geers, "Thermomechanical fatigue damage evolution in SAC solder joints," Materials Science and Engineering, A, doi: 10, 1016/j.msea. 9, 2006, 37.
9. L. L. Ye, Z. Lai, J. Liu, A. Thölén, "Microstructural coarsening of lead free solder joints during thermal cycling," IEEE, Electronic Components and Technology Conference, Las Vegas, NV, 2000, 134–137.
10. P. Towashiraporn, K. Gall, G. Subbarayan, B. McIlvanie, B. C. Hunter, D. Love, B. Sullivan, "Power cycling thermal fatigue of Sn–Pb solder joints on a chip scale package," International Journal of Fatigue, 26(5), 2004, 497–510.
11. P. G. Harris, K. S. Chaggar, M. A. Whitmore, "The effect of ageing on the microstructure of tin–lead alloys," Soldering and Surface Mount Technology, 7, 1991, 24–33.
12. R. Agarwal, S. E. Ou, K. N. Tu, "Electromigration and critical product in eutectic SnPb solder lines at 100°C," Journal of Applied Physics, 100(024909), 2006, 1–5.
13. H. Ye, C. Basaran, D. C. Hopkins, "Pb phase coarsening in eutectic Pb/Sn flip chip solder joints under electric current stressing," International Journal of Solids and Structures, 41, 2004, 2743–2755.
14. H. L. J. Pang, K. H. Tan, X. Q. Shi, Z. P. Wang, "Microstructure and intermetallic growth effects on shear and fatigue strength of solder joints subjected to thermal cycling aging," Materials Science and Engineering, A, 307, 2001, 42–50.
15. J. W. Jang, A. De Silva, J. K. Lin, D. Frear, "Mechanical tensile fracture behaviors of solid-state-annealed eutectic SnPb and lead-free solder flip chip bumps," IEEE Electronic Components and Technology Conference, 2003, 680–684.
16. R. Erich, R. J. Coyle, G. M. Wenger, A. Primavera, "Shear testing and failure mode analysis for evaluating BGA ball attachment," IEEE/CPMT, International Electronics Manufacturing Technology Symposium, Austin, TX, 1999, 16–22.
17. S. W. Kim, J. W. Yoon, S. B. Jung, "Interfacial reactions and shear strengths between Sn–Ag based Pb-free solder balls and Au/EN/Cu metallization," Journal of Electronic Materials, 33(10), 2004, 1182–1189.
18. A. Sharif, Y. C. Chan, M. N. Islam, M. J. Rizvi, "Dissolution of electroless Ni metallization by lead-free solder alloys," Journal of Alloys and Compounds, 388(1), 2005, 75–82.
19. P. L. Liu, J. K. Shang, "Influence of microstructure on fatigue crack growth behavior of Sn-Ag solder interface," Journal of Electronic Materials, 29(5), 2000, 622–627.
20. K. S. Kim, S. H. Huh, K. Suganuma, "Effects of intermetallic compounds on properties of Sn–Ag–Cu lead-free soldered joints," Journal of Alloys and Compounds, 352, 2003, 226–236.
21. Q. Xiao, H. J. Bailey, W. D. Amstrong, "Aging effects on microstructure and tensile property of Sn3.9Ag0.6Cu solder alloy," Journal of Electronic Packaging, 126(2), 2004, 208–212.
22. B. F. Dyson, "Diffusion of gold and silver in tin single crystals," Journal of Applied Physics, 37, 1966, 2375–2377.
23. Y. Kariya, M. Otsuka, "Effect of ismuth on the isothermal fatigue properties of Sn-3.5 mass% Ag solder alloy," Journal of Electronic Materials, 27(7), 1998, 866–870.
24. L. Qi, J. Zhao, X. M. Wang, L. Ang, "The effect of Bi on the IMC growth in Sn-3Ag-0.5Cu solder interface during aging process," International Conference on Business of Electronic Product Reliability and Liability, 2004, 2–46.
25. C. M. L. Wu, M. L. Huang, "Microstructural evolution of lead-free Sn–Bi–Ag–Cu SMT joints during aging," IEEE Transactions on Advanced Packaging, 28(1), 2005, 128–133.
26. K. S. Kim, S. H. Hug, K. Suganuma, "Effect of fourth alloying additive on microstructures and tensile properties of Sn–Ag–Cu alloy and joints with Cu," Microelectronics Reliability, 43, 2003, 259–267.
27. P. Sun, C. Andersson, X. Wei, Z. Cheng, D. Shangguan, J. Liu, "Intermetallic compounds formation in Sn–Co–Cu, Sn–Ag.Cu and eutectic Sn–Cu solder joints on electroless Ni(P) surface finish after reflow soldering," Materials Science and Engineering, B, 135, 2006, 134–140.

28. S. Dunford, S. Canumalla, P. Viswanadham, "Intermetallic Morphology and Damage Evolution Under Thermomechanical Fatigue of Lead(Pb)-Free Solder Interconnections," Proc. IEEE Electronic Components and Technology Conference, 2004, 726–736.
29. K. Norris, A. Landzberg, "Reliability of Controlled Collapse Interconnections," IBM J. Res. Dev. Interconnection Reliability, 1969, 266–271.
30. N. Pan, G. Henshall, F. Billaut, S. Dai, M. Strum, R. Lewis, E. Benedetto, J. Rayner, "An acceleration model for Sn–Ag–Cu solder joint reliability under various thermal cycle conditions," Proc. SMTAI, 2005, 876–883.
31. O. Salmela, "Acceleration factors for lead-free solder materials," Transactions of IEEE Components & Packaging Technologies, 30(4), 2007, 700–707.
32. R. Wassink, M. Verguld, "Manufacturing Techniques for Surface Mounted Assemblies," Electrochemical Publications LTD, Bristol, England, 1995, 83–85.
33. J. Hwang, "Environment-Friendly Electronics: Lead-Free Technology," Electrochemical publications LTD, Bristol, England, 2001, 65–69.
34. O. Vianco, "Corrosion Issues in Solder Joint Dedsign and Service," Energy Citations Database, 1999, http://www.osti.gov/energycitations/purl.cover.jsp?purl=/14961-8gkDYa/webviewable/
35. T. Mattila, P. Marjamäki, J. Kivilahti, "Reliability of CSP Interconnections Under Mechanical Shock Loading Conditions," IEEE Transactions on Components and Packaging Technologies, 2006, 29(4), 787–795.
36. A. Prabhu, W. Schaefer, S. Patil, "High Reliability LTCC BGA for Telecom Applications," Proc. IEEE International Electronics Manufacturing Technology Symposium, 2000, 311–323.
37. "RIAC Handbook of 217PlusTM Reliability Prediction Models," Department of Defense, USA, 2006, 145–147.

Chapter 5
Conductive Adhesive Joint Reliability

Abstract There are two primary categories of electrically conductive adhesive (ECA): isotropic conductive adhesive (ICA) and anisotropic conductive adhesive (ACA), where ACAs are available as paste (ACP) or film (ACF). Both types conduct through metal filler particles in an adhesive polymer matrix.

This chapter presents an overview of the current status of understanding of conductive adhesives in various electronic packaging applications and of some fundamental issues relevant to their continuing development. It is organized with initial discussions of basic ECA concepts of structure-related properties, and how these are affected by material selection and processing, followed by general properties and reliability considerations.

5.1 Introduction to Conductive Adhesives

Recent environmental legislation has led to an increasing interest in the possibility of substituting electrically conductive adhesives (ECAs) for the traditional tin–lead solders in electronics manufacturing. The conductive adhesives mentioned in this chapter are not inherently conductive polymers that are extremely brittle and sensitive to oxidation. Instead, they are composites of insulating polymer matrix and conductive fillers. The polymer matrix and its characteristics are mostly responsible for the adhesive ability to bond and withstand mechanical stresses. The electrical conductivity of the adhesive depends particularly on the fillers. As a result, the electrical and mechanical properties can, to a large degree, be adjusted independently. Depending on the loading of fillers, conductive adhesives can be cataloged as isotropic conductive adhesive (ICA) and anisotropic conductive adhesive (ACA). Though not every adhesive currently has all of them, conductive adhesive interconnections offer the following advantages over traditional tin–lead solders:

- Low temperature processing
- Compatibility with a wide range of substrates
- No flux pretreatment or postcleaning procedures required

J. Liu et al., *Reliability of Microtechnology: Interconnects, Devices and Systems*,
DOI 10.1007/978-1-4419-5760-3_5, © Springer Science+Business Media, LLC 2011

- No lead or other toxic metals
- Finer pitch capability
- Solder mask not required

5.2 Isotropic Conductive Adhesive

ICAs have been successfully used for decades in the electronics industry as die-attach materials. Now new adhesives have been formulated to replace traditional solders in mainstream applications. The volume fraction of conductive fillers in the adhesive is between 20 and 35%, which is so high that the adhesive can conduct equally well in all directions. As a result, ICAs may be deposited only where electrical connects are required. In general, the conductivity of the adhesive improves with increasing filler loading, but at the expense of the adhesive becoming increasingly brittle. Copper, nickel, carbon, and silver are commonly used as conductive fillers. Silver is unique among these affordable fillers because of its good electrical performance, stability, and inherent conductivity of silver oxides. The matrix is mostly one- or two-component epoxies that can be cured with heat and/or IR radiation. However, polyimides, silicones, and thermoplastic adhesives can also be used as matrices.

The electrical conduction of an ICA joint is primarily established during cure. Instead of metallurgical connection, the joint conduction is based on mechanical contacts among conductive fillers (Fig. 5.1). Studies have shown that the conduction development during cure is accompanied by the decomposition of organic lubricants, which exposes the metallic surface of fillers, and the cure shrinkage, which brings fillers closer. However, the conduction mechanism of ICA is still not fully understood and which effect plays a dominant role is still open to question.

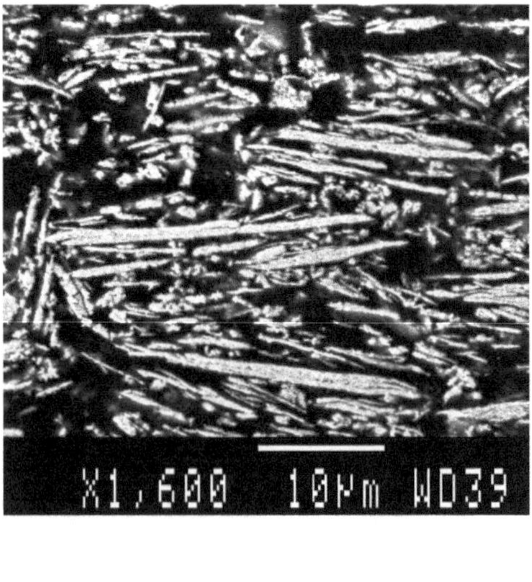

Fig. 5.1 Microstructure of an ICA showing silver fillers (*white*) embedded in the epoxy matrix (*black*)

After the adhesive is cured, the fillers are randomly distributed and form a network within the polymer matrix. By this network, electrons can flow from one adherent to the other across the filler contact points. The overall result is to create numerous electron pathways, but with each path made up of a large number of mechanical contacts. So any factors affecting intimate contacts among fillers will surely influence the performance and reliability of ICA interconnects.

Besides die attachment, ICAs are utilized in surface mount and flip-chip packages as alternatives to traditional solders. But, due to their low surface tensions, ICAs are not suitable for wave soldering. Despite the advantages of ICA interconnection, the wide use of this technology has not been adopted by the electronics industry. The main concern is the long-term reliability.

5.3 Reliability of ICA Interconnects

5.3.1 Effect of Metallization

To get a good adhesive joint, the adhesive must wet the bonding surface. A necessary condition for this is that the adhesive has lower surface tension than the bonding surface. Epoxy and polyimides are major polymers used as base matrices of ICAs. These materials have lower surface tension than Sn, Pb, Cu, Au, and Pd. Therefore, a good adhesive joint is expected when bonding on Sn, SnPb, Cu, AgPd, and Au surfaces.

As water molecules can easily penetrate through the adhesive and oxidize/hydrate the bonding surfaces, ICA joints with different metallizations have diverse reliability performances in high-humidity environments. Several investigations showed that joints with noble Au and AgPd metallizations had much less resistance increases compared with those with non-noble SnPb and Cu metallizations. The detailed failure mechanisms were investigated with transmission electron microscopy (TEM) and X-ray electron spectroscopy for chemical analysis (XPS/ESCA).

TEM observations on an adhesive joint with Sn37Pb metallization show that water has penetrated to the Sn37Pb surface after 1,000 h 85°C/85%RH test. As a result, oxygen signals can be detected with EDS analysis in TEM. The corresponding electron diffraction observes a diffused ring, which indicates that Pb was converted to an amorphous structure. So the reaction product is not crystalline PbO, but $Pb(OH)_2$ or other Pb oxides such as Pb_2O_3 and Pb_2O which have amorphous structures. The ESCA analysis on the Sn37Pb surface shows the chemical shift of the oxygen signals, confirming that the product is Pb-hydroxide. Due to the formation of amorphous $Pb(OH)_2$ which is an insulating compound and has a powdery structure, both electrical and mechanical properties of the ICA joint deteriorate. With the same metallization, Botter et al. [1] and Jagt [2] got similar resistance shift trends. But they focused more on tin oxidation according to electrochemical analysis.

TEM studies on an adhesive joint with copper metallization show the existence of the oxide layer on the Cu pad. The thickness is approximately 100 nm after 1,000 h 85°C/85%RH humidity test. The rings in the diffraction pattern obtained from the oxide layer indicate that the layer consists of fine crystalline grains. The radii of the rings correspond to the spacing of the crystallographic planes of Cu_2O. It is, therefore, concluded that the formed oxide is Cu_2O, which is a poor conductor. This helps to explain why the joint resistance increased after the humidity test.

Contrary to non-noble metallization, the electrical resistance of ICA joints mounting a gold-plated QFP80 component on an electroless Au-plated FR-4 board is quite stable in the 85°C/85% RH environment. No significant increase can be observed up to 2,000 h. Hence, it can be concluded that a noble metallization such as Au or Ag/Pd is preferable for normal ICAs. However, by adding corrosion inhibitors, some superior ICAs for pretinned metallization have been developed.

5.3.2 *Effect of Curing Degree*

There is no doubt that proper curing is very important for joint reliability. It was found that a minimum curing degree is required to provide a certain level of mechanical and electrical performance of adhesive joints, especially with non-noble metallizations. Once this is achieved, increasing curing times does not result in significant improvement.

Figure 5.2a shows the electrical resistance shifts of epoxy-based ICA joints after 1,000 h humidity test at 85°C/85%RH. These joints were on the Sn37Pb bonding surface and cured at 150°C for various time. The corresponding curing degrees were determined by differential scanning calorimetry (DSC) measurement as between 65 and 90%. Below a critical curing degree (for this adhesive, the critical curing degree is 77%), the electrical resistance of the joint increases significantly after humidity test. The reason is that an undercured epoxy can absorb a significant amount of water, which in turn causes oxidation/hydration of the Sn37Pb metallization. If a noble metallization such as AgPd is used, no electrical resistance shift

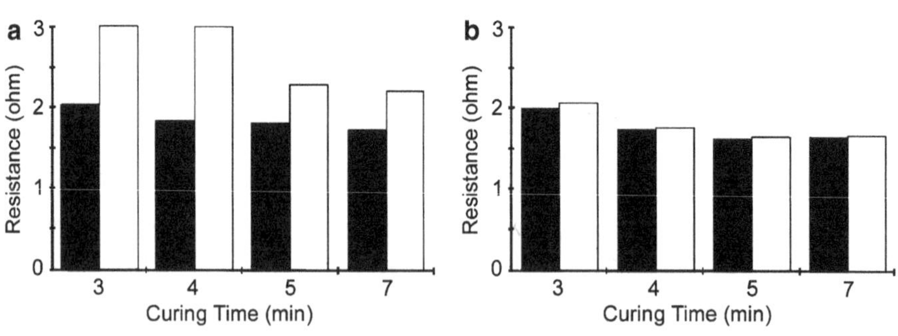

Fig. 5.2 Contact resistance shifts of ICA joints on (**a**) Sn37Pb and (**b**) Ag/Pd surfaces (*black*: before test; *white*: after test)

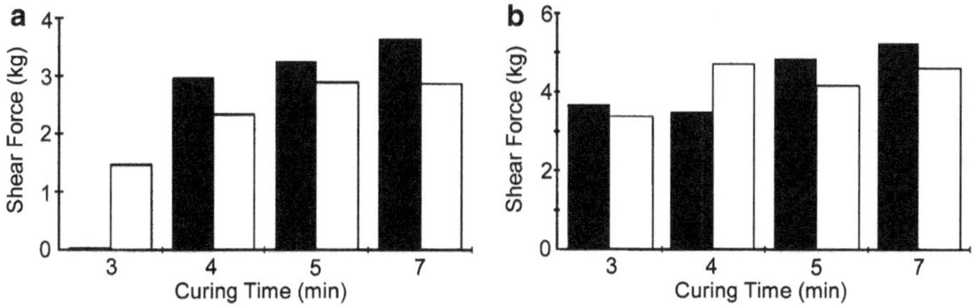

Fig. 5.3 Strength shifts of ICA joints on (**a**) Sn37Pb and (**b**) Ag/Pd surfaces (*black*: before test; *white*: after test). These joints were cured at 150°C for various time and then aged in the 85°C/85% RH environment

has been observed despite the fact that curing degree can be very low, as can be seen in Fig. 5.2b. These joints were cured at 150°C for various times and then aged in the 85°C/85%RH environment.

The same as electrical performance, once a critical curing degree is achieved (77%), the shear strength of the joint on the Sn37Pb bonding surface can be maintained at a constant level (Fig. 5.3a). However, on the noble metal bonding surface, the shear strength of the joint is almost independent of the curing degree in the range between 65 and 90%, as can be seen in Fig. 5.3b. These results also indicate that for conductive adhesive joining, noble metallization is preferable to non-noble metallization.

5.3.3 Impact Strength

Due to their high filler loading, many ICAs suffer from poor impact strength, which is one of the major drawbacks preventing their wide applications. Without adequate impact strength, ICA joints can hardly survive the significant shocks during assembly, handling, and usage. For bulk materials, impact performance is closely related to their fracture toughness and damping property. An adhesive with higher toughness and higher loss modulus normally has better impact performance. So a simple approach is to modify base epoxy resins with elastomers to improve the impact performance of ICAs. However, for adhesive joints, the adhesion strength between the adhesive and adherend is also very critical. Low impact strength can result from adhesive failure due to poor adhesion. Using conformal coating of surface mount devices is another practical way to improve the impact strength of the package.

To evaluate the impact performance of board-level packaging with ICAs, the National Center for Manufacturing Science (NCMS) has developed a special drop test. It involves dropping circuit boards onto hard ground from a height of 1.5 m and the sample surviving six drops is regarded to possess acceptable impact strength.

The test is easy to conduct, but only qualitative information could be supplied. Xu and Dillard [3] developed a novel falling wedge fracture test which is capable of quantitatively determining the impact strength of ICA joints. They used a modified double cantilever beam (DCB) specimen with ICA and PCB boards and measured the fracture energies under different test temperatures with this new test technique. The impact fracture energy was found to keep an approximately logarithmic relationship with the loss factor that in turn can serve as a good indicator of the impact performance of ICAs.

5.3.4 Failure Mechanisms

5.3.4.1 Cracking

Due to the temperature fluctuation caused by the circuit power on/off cycles, ICA interconnects have to sustain cyclic stresses from thermal expansion mismatch between the substrate and component, and thermomechanical fatigue cracking is considered as one of the primary failure mechanisms. Based on temperature cycling tests and cross-section observations, the fatigue cracking behavior of ICA joints of leadless chip resistors was investigated. Early cracking was detected at the top of the vertical adhesive/termination interface (Fig. 5.4a), which has been reported in [4]. With more cycles, cracks were observed at the inner end of the horizontal interface between adhesive and ceramic resistor body (Fig. 5.4b). As the number of cycles increased further, bulk cracking occurred around the knee of the joint (Fig. 5.4c). It appears that several microcracks nucleated simultaneously due to the debonding of silver flakes. Then they merged together and formed the main bulk crack that propagated from the component side to the board side. After initiation, both vertical and horizontal cracks propagated toward the knee area along the adhesive/termination interface and the final merging of the three cracks (Fig. 5.4d) resulted in a complete failure of the entire joint. Since most crack development occurs at the interface, the adhesion of ICA is critical to the joint reliability.

In humidity aging tests, cracks have always been found associated with electrical degradation of ICA joints. Li et al. [5] reported that cracks existed after cure and developed due to the humidity exposure, leading to deterioration in both mechanical strength and electrical conductivity. However, with similar observations, Botter et al. [1] attributed the cracking after humidity test to the formation of oxides. In a recent investigation on the degradation of ICA joints in humid environments, Xu et al. [6] concluded that moisture attack on the adhesive/metallization interface could be divided into three phases: displacing the adhesive due to high surface-free energy around the interface, hydrating the metal or metal oxide, and forming a weak boundary layer at the interface. If the attack occurs in the first phase, the fracture energy could recover to some extent after redrying at high temperature. However, the degradation becomes irreversible in the second phase.

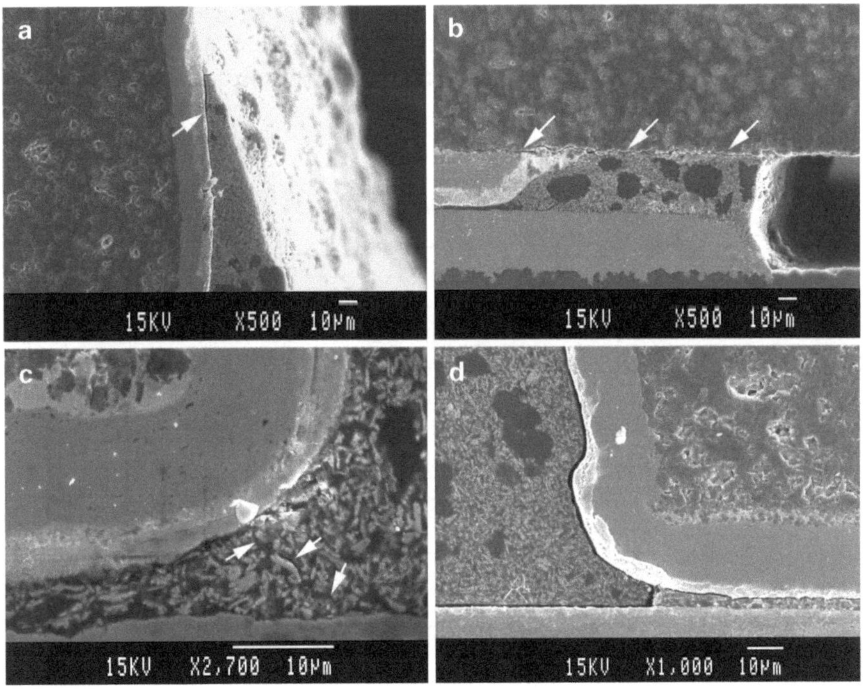

Fig. 5.4 Interfacial cracks initiated at (**a**) top and (**b**) inner ends of the adhesive/component interface; (**c**) bulk microcracks occurred around the knee of the joint. Cracks are indicated with *arrows*; (**d**) final merging of these cracks resulted in a complete failure of the entire joint

5.3.4.2 Formation of Oxides

When ICAs are used together with non-noble metallizations, the contact resistance will increase significantly during high temperature and high humidity aging. As discussed in Sect. 4.3.2, various oxides have been observed to form at the interface between ICA and metallization, which leads to resistance deterioration.

In climate tests (98%RH), Botter et al. found that tin oxides occurred only in the area where adhesive was attached. No visible oxidation was observed at the fully air-exposed area on the top of the resistor. So they proposed that the direct contact between the noble metal (Ag filler) and the non-noble metal (tin metallization), combined with absorbed water in the adhesive, formed a local electrochemical cell, which corroded the non-noble metal. They also pointed out that tin oxide has no passivation effect and the contact resistance would increase progressively when tin is present in the metallization. However, metallizations of pure Cu and Pb were acceptable in high humid environments because oxides of Cu and Pb tend to form dense layers and further corrosion can be hindered.

Lu et al. [7] showed additional evidences supporting the electrochemical corrosion mechanism. They found that the joint resistance could keep stable either in dry environments or under the 85°C/85%RH condition but only one metal was involved.

Only if two different metals (e.g., Ni fillers and Ag wire) were involved, would the joint resistance increase dramatically. By formulating low moisture absorption resin and adding corrosion inhibitors, these authors succeeded in developing high-performance ICAs.

5.3.4.3 Formation of Intermetallic Compounds

Increase of resistance after environmental tests can also be attributed to the formation of intermetallic compounds. Yamashita and Suganuma [8] investigated the heat-induced degradation of the interface between ICA and SnPb-plated Cu electrode. Their element mapping analysis showed the apparent Sn diffusion into Ag particles. The occurrence of Ag–Sn intermetallic compounds, such as Ag_3Sn and Ag_4Sn, was identified in the X-ray diffraction pattern. They attributed this phenomenon to the Kirkendall diffusion of Sn from the plating layer into the Ag particles. At 150°C, the diffusion constant of Sn in Ag ($2.31 \times 10^{-17} m^2/s$) is much larger than that of Ag in Sn ($2.32 \times 10^{-20} m^2/s$). Therefore, the preferential diffusion of Sn occurs. This results in large Kirkendall voids in the SnPb plating layer, which decreases the true bonding area of the ICA joint and thus degrades both electrical and mechanical properties. These authors also pointed out that the diffusion constant of Sn in Ag_3Sn ($6.37 \times 10^{-12} m^2/s$) is even higher than that of Sn in Ag and the formation of Ag_3Sn cannot hinder the Kirkendall diffusion of Sn.

5.3.4.4 Filler Motion

Several researchers [9–12] have noticed the difference in deformation behaviors of metal fillers and polymer matrix. Typically conductive adhesive joints can sustain a shear strain of 10%, which is an order great than solders. But the metal fillers in ICA cannot be strained that much. Instead, they would move relatively to one another due to the compliancy of the matrix. Some possible influences on ICA reliability were proposed, concerning this situation.

Keusseyan et al. [9] observed that compliant adhesive joints could survive more than 3,000 thermal cycles without losing much mechanical strength, but the electrical resistance increased significantly. They suggested that relative movement among fillers, combined with viscoplastic deformation of matrix, would pull the insulating polymer in between fillers, leading to the loss of interfiller contacts.

With similar observations, Rörgren and Liu [10] suggested that the filler motion would result in sliding along the interface between fillers. When the adhesive joint is subject to cyclic loadings, this interfacial sliding would eventually wear out the direct contact points among fillers and degrade the electrical performance of the ICA joint in the long run. Besides filler friction, the numeric simulation [11] showed that stress concentration due to filler motion would promote the initiation of microcracks in polymer matrix, which could weaken the constraint on fillers, loosen their intimate contacts, and therefore increase the bulk resistance.

Constable et al. [12] performed mechanical low cycle fatigue tests on several ICA joints and measured the resistance changes with a highly sensitive micro-ohm technique. The resistance was observed to increase apparently at the initiate stage of the tests, while the force required for the same deformation amplitudes decreased gradually. The authors attributed this phenomenon to the formation of wear tracks from filler frictions. However, they insisted that the influence of filler motion is limited and the dominant failure mechanism is interfacial fracture of the joint.

5.3.4.5 Ag Migration

In the presence of water and an electric field, silver is anodically dissolved at its original location and moves toward the cathode where it is deposited. This migration phenomenon can lead to the growth of dendrites between adjacent electrodes and lower the surface insulation resistance (SIR) of the board. For many years, the short circuit due to Ag migration has been a nuisance to those using silver inks and similar.

However, due to that the silver fillers are encapsulated with an epoxy layer, Ag migration is not likely to occur in conductive adhesives under test conditions relevant in practice, e.g., 85°C/85%RH or 60°C/90%RH under 5 V bias [2]. But under more severe conditions, such as the presence of a liquid water film, higher bias and smaller pitch spacing, Ag migration does occur. For example, short circuit between 8-mil spaced pads has been observed after 2,000 h of 85°C/85%RH test with 15 V bias (Fig. 5.5). In ref. [13], the migration of Ag particles was also observed in ICA joints subjected to the current-induced aging (10–30 A) and the consequent electrical degradation was reported.

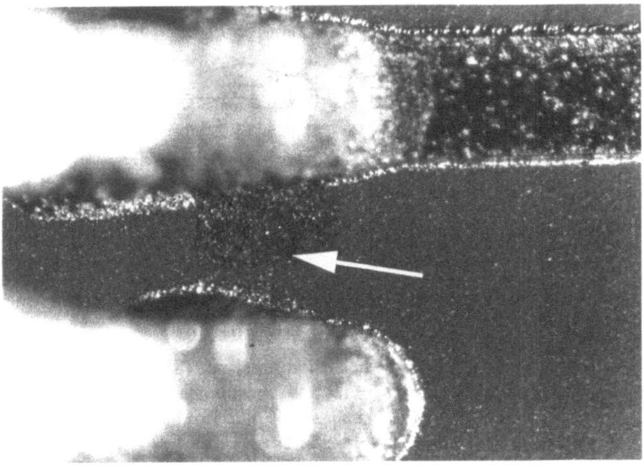

Fig. 5.5 Ag migration between 8-mil spaced pads after 2,000 h of 85°C/85%RH test with 15 V bias

5.3.5 Electron Conduction Through Nanoparticles in ICA

High metal loading, in the range of 20–35 vol% is normally required to guarantee effective electrical conduction of ICA joints, which results rather often in adhesive failure. Based on the percolation theory, ICAs using a bimodal distribution of metal fillers were expected to have a decreased metal loading for better mechanical performance while the electrical property remains unchanged [14]. It was, however, demonstrated experimentally that the electrical conductivity has been reduced when the volume percentage of the nanosize fillers in the system was increased [15]. To explain this phenomenon, the electron conduction through nanoparticles in a normal ICA was investigated based on quantum-mechanical considerations [16].

Consider a substructure in the ICA that consists of one nanoparticle sandwiched between two microparticles (Fig. 5.6).

For the nanoparticle between the two microparticles, the quantum confinement effects must be included. Using the uniform background model, the ground sublevels across the substructure are approximately:

$$E_{nm}(z) = \frac{\pi^2 \hbar}{2r(z)^2 m_0}(n^2 + m^2), \tag{5.1}$$

where $r(z)$ is the radius of the cross section at z, n and m are nonzero integers. The ground sublevel $E_{11}(z)$ is also presented in Fig. 5.6, where a potential barrier in the nanoparticle side of the interconnect ($z = 0$) between the micro- and nanoparticles is observed.

When the structure is biased by V_{ex}, the local Fermi level of the left microparticle is kept unchanged (assumed to be grounded). The local Fermi level of the other microparticle becomes $E_f + eV_{ex}$, its conduction band edge E_{11} is also lifted up by an amount of eV_{ex}. The time-dependent quantum mechanical behavior of an electron can be described by its wave packet. As the electron transports through the substructure of Fig. 5.6 from left to right, the electron wave packet is split into two parts after

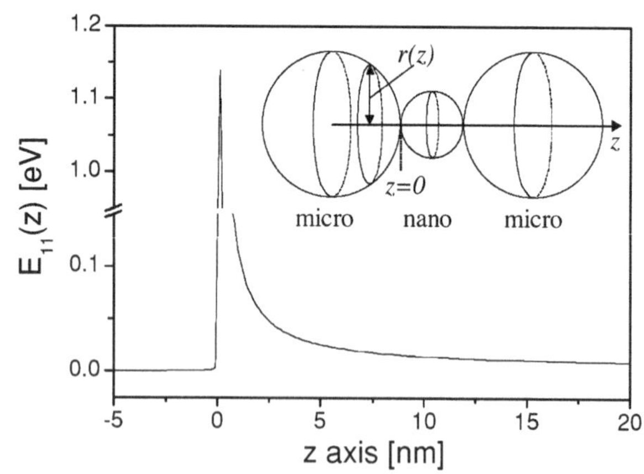

Fig. 5.6 Schematics of a micro–nano–micro substructure in ICA and the potential energy profile at the interconnect between the micro- and nanoparticles

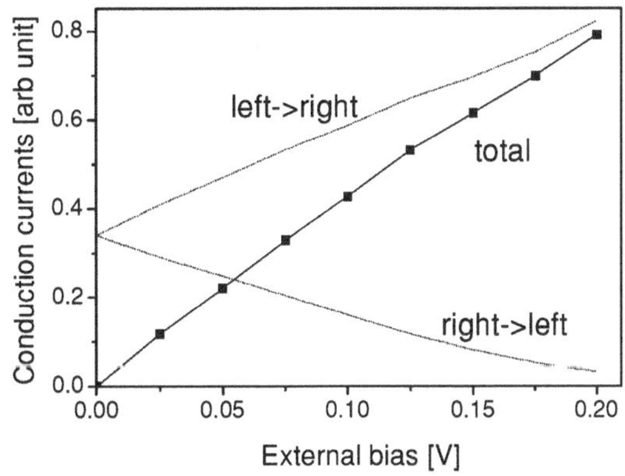

Fig. 5.7 Current–voltage characteristics of the micro–nano–micro substructure

reaching the left interconnect energy barrier at $z = 0$. One is reflected back, and the other tunnels through the nanoparticle. Calculation shows that, due to the barriers, only half of the initial electron gets transmitted through the nanoparticle. This qualitatively agrees with experimental observations that the electrical conductivity has been reduced as nanosize fillers are added into the ICA.

The current–voltage characteristics of such a structure are presented in Fig. 5.7. The total current is obtained by subtracting the reflected current from the transmitted current. Increasing the external bias effectively lowers the energy barriers so that the transmitted current increases; however, it lifts up the conduction band edge E_{11} of the left microfiller so that the reflected current decreases considerably. The final total current through the substructure in general increases linearly in the external bias range under investigation.

5.4 Reliability of ACA Interconnects

Recent environmental legislation has led to an increasing interest in the possibility of substituting ECAs for the traditional tin–lead solders in electronics manufacturing. The conductive adhesives mentioned in this chapter are composites of insulating polymer matrix and conductive fillers. Depending on the loading of fillers, conductive adhesives can be cataloged as ICA and ACA. In this section, our topic focuses on the ACA. ACA is a new class of adhesives that are conductive in one direction, which offers the following advantages over traditional tin–lead solders in the interconnections.

- Low-temperature processing
- Compatibility with a wide range of substrates
- No flux pretreatment or postcleaning procedures required
- No lead or other toxic metals
- Finer pitch capability
- Solder mask not required

ACAs are prepared by dispersing conductive fillers in an adhesive matrix. The unidirectional conduction is achieved by using a relatively low volume fraction of conductive fillers. This low filler loading is insufficient for interparticle contact and prevents conduction in the *X–Y* plane of the adhesive, but enough particles are present to assure reliable conduction between bonding electrodes in the *Z*-direction. Because of the anisotropy, ACAs can be deposited over the entire contact region, greatly expanding the bonding area. Also the low filler loading improves the bonding strength. Thus, mechanically robust interconnection can be achieved with ACA assembly.

ACAs come in two distinct forms: paste and film. Pastes can be printed with screen or stencil or dispensed with a syringe. Films are supplied by manufacturers on reel and are extremely suitable for nonplanar bonding surfaces. Both thermoplastic and thermosetting resins have been used as adhesive matrices. The principal advantage of thermoplastic ACAs is the relative ease to disassemble the interconnections for repair operation, while thermosetting adhesives possess higher strength at elevated temperature and form more robust bonds [17]. The commonly used conductive fillers include silver and nickel particles and polymer spheres coated with metal (Ni/Au). Silver particles offer moderate cost, high electrical conductivity, and low chemical reactivity. Nickel particles can break the oxide layer on the electrodes and are suitable for interconnecting easily oxidized metal. Metal-coated polymer spheres have fairly uniform diameter distributions. They can provide high interconnection reliability because of the large elastic deformation during bonding. Recent application of solder particles as ACA fillers has also been reported [18].

Since the conduction of ACAs is based on mechanical particle–electrode contacts, pressure is a requisite to form qualified joints. A typical ACA assembly is shown in Fig. 5.8. After alignment, pressure is applied on the backside of the chip. The adhesive resin is squeezed out and conductive particles are trapped and deformed between opposing electrodes. Once electrical continuity is generated, the adhesive resin is cured with heat or UV. The intimate particle–electrode contacts are maintained by the cured matrix and the elastic deformation of particles, and electrodes exert a continuous contact pressure.

ACA interconnection finds particular applications with fine-pitched flip-chip techniques used to mount bare chip on various substrates such as ITO-coated

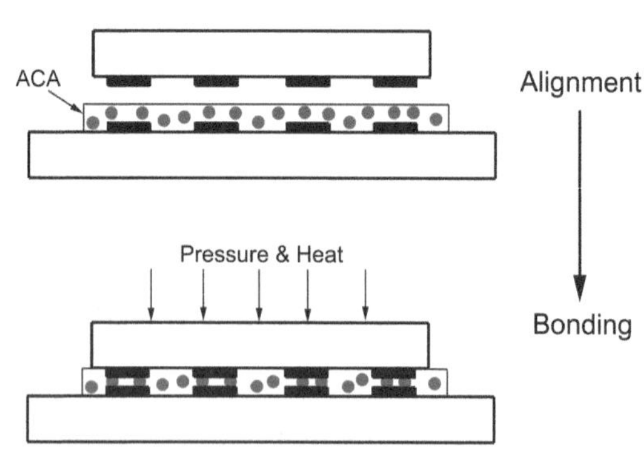

Fig. 5.8 Manufacturing process of ACA assembly

glass, FR4 board, and flexible films. ACA joining is also attractive for fine-pitched surface mount component assembly. However, the performance and reliability of ACA joints are more sensitive to the joint design, substrate/component properties, and process conditions than solder joints.

5.4.1 Effects of Assembly Process

The assembly process of ACA interconnection includes alignment, bonding and, if solder interconnects exist in the same board, reflow. Due to the low surface tension, ACA interconnection lacks the benefit of the self-alignment, which put a stringent requirement on the alignment accuracy. A normal flip-chip bonder that offers a $\pm 5\ \mu m$ accuracy is normally good enough. Nevertheless, bad alignment would result from incorrect operations. It can influence the pressure distribution and, in more serious situations, decrease the contact area for electrical interconnection (Fig. 5.9).

The bonding process is very critical to the ACA joint performance and reliability, since both mechanical integration and electrical interconnection are established in this process. Bonding pressure and temperature are the two most important parameters. To achieve reliable ACA joints, adequate bonding pressure should be applied uniformly and suitable bonding temperature should be kept for sufficient time.

The bonding pressure is applied to force the conductive particles to contact the electrodes. The performance of the joint depends heavily on the deformation degree of particles. Ideally, the particles should be squashed enough to gain the largest contact area. However, the integration of particle body should be maintained and cracking due to over pressure could degrade the electrical performance.

Fig. 5.9 Bad alignment degrades the electrical performance and reliability of ACA joints

It is also important to keep the pressure uniform during the bonding. Nonhomogeneous bonding pressure can cause particles being deformed unevenly, which could result in poor long-term reliability. This problem becomes more serious for thin and flexible substrates.

The effects of particle deformation on joint electrical reliability during temperature cycling are summarized schematically in Fig. 5.10. Type 1 represents the best case where the particles are deformed uniformly and atomic bonding between the particles and contacts is achieved. Type 2 joints consist of undeformed or slightly deformed particles due to either low bonding pressure or inhomogeneous pressure distribution. The conductive character of these joints is unstable at high temperature because the epoxy matrix will expand more than the particles. Type 3 joints can

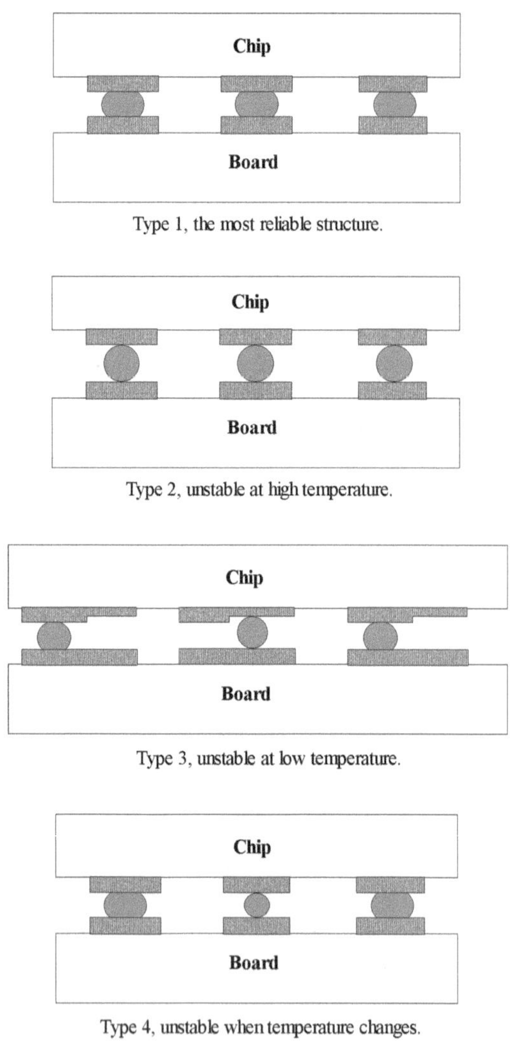

Fig. 5.10 Schematics of four types of ACA joints caused by variations in bonding pressure, bump geometry, and filler size

result from shape or height variations of the contact areas. Some particles are not deformed enough and will shrink more than those well deformed, causing problems at low temperature. Finally, type 4 pictures a uniform height of the contact areas, but a very large variation of particle size. Due to the weak bonding between the smaller particles and contact area, electrical opens can be observed at both low and high temperatures. All these situations have been observed experimentally.

The bonding temperature and time heavily influence the curing degree of the adhesive that plays an important role in the reliability of ACA joints. In the under-cured joint, the cross-linkage of the polymer may be incomplete and neither mechanical performance nor electrical reliability can be guaranteed under high humidity tests. To gain a certain curing degree, longer bonding time should be employed with lower bonding temperature. However, this is not preferable due to the low productivity. On the other hand, too high bonding temperature is not desired, either. This is because the epoxy may solidify too quickly and hence the conductive particles would not have enough time to distribute themselves in between the bumps and pads. Recent work also observed the chain scission due to high bonding temperature. So finding the optimum combination of bonding temperature and time is a fundamental step toward reliable ACA interconnection.

If ACA interconnection is used together with soldering technology for the final products, reflow soldering after ACA bonding is inevitable. During reflow, the package needs to be heated up to above 200°C, which is much higher than the normal bonding temperature of ACA joint. The ability of ACA interconnection to withstand this high temperature is critical for successful packaging. Yin et al. [19] found that the contact resistance of ACA joints increased significantly after reflow process and conduction gaps formed between the conductive particles and the electrode. Seppälä and Ristolainen [1] also reported the detrimental effects of reflow on the reliability of ACA joints. The possible reason is that, due to its much higher coefficient of thermal expansion (CTE), the adhesive matrix expands in the *Z*-direction much more than the particles during the reflow. The induced thermal stress lifts the chip from substrate and damages the bonding structure. Therefore, the peak temperature of reflow profile and the distance between the chip and substrate (related to bump height) are the most important factors. By optimizing process parameters and adopting ACA with lower CTE, the effects of the reflow process can be reduced to some extent.

5.4.2 Effects of Substrate and Component

Suitable substrate stiffness and bump dimensions are also important to achieve reliable ACA joints. With a soft substrate, significant deformation of the substrate may occur during the bonding, which has a direct influence on the joint quality. On the FR4 board, it was observed that the electrical resistance and reliability of a joint depend on the distance between the pad and glass fibers in the substrate (Fig. 5.11). A long distance means a thick layer of soft epoxy that may deform during bonding. Therefore, enough particle deformation cannot be obtained at that point.

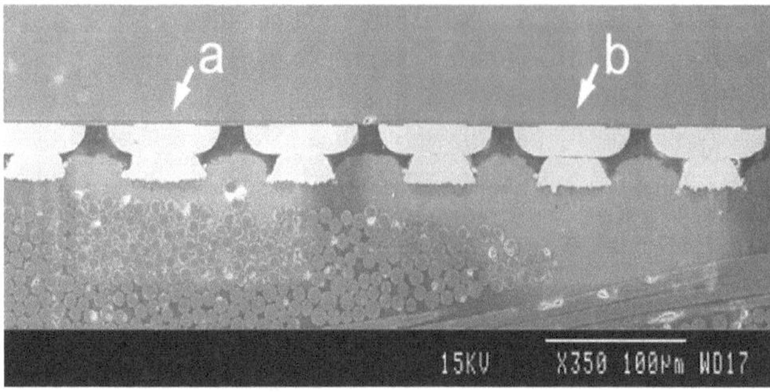

Fig. 5.11 Electrical resistance and reliability of a joint depend on the distance between the pad and glass fibers in the substrate. Joint **a** has a better electrical performance than joint **b** (5 vs. 14 mΩ) due to its location closer to glass fibers

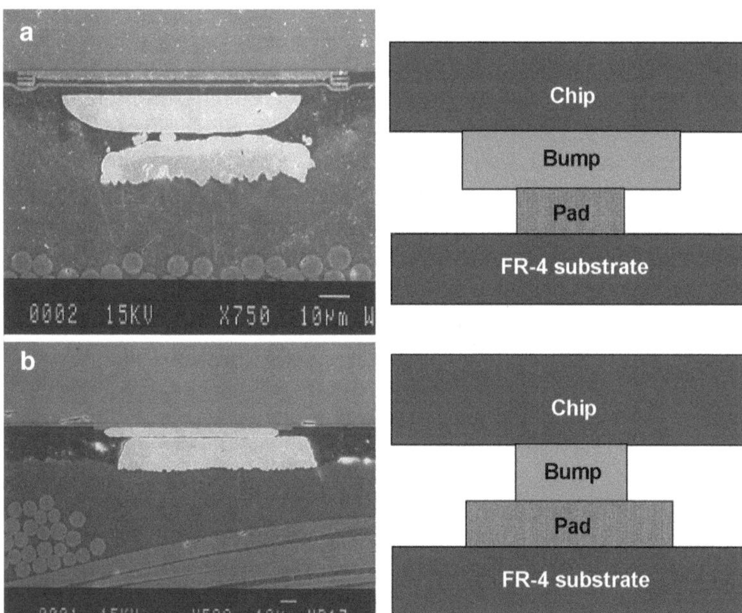

Fig. 5.12 (**a**) Pad sinking leads to insufficient particle deformation and (**b**) using a bump smaller than the pad can decrease pad sinking

Figure 5.12a shows that large force exerted on the pad causes the pad sinking and almost no deformation occurring in the particles. An approach to reduce the pad sinking is to use a relatively smaller bump area compared with the pad area. Therefore, less bonding force will be transferred to the pad as shown in Fig. 5.12b.

For flip-chip solder joining, plastic strain of solder bumps is a critical parameter that governs the joint reliability. Using a high bump can reduce the bump strain and thus increase the joint reliability, as shown in Fig. 5.13a. However, a systematic

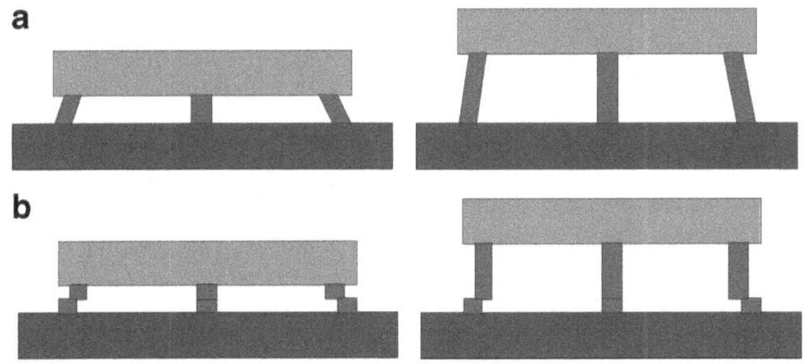

Fig. 5.13 (**a**) Using high bumps can reduce the strain of solder joints, but (**b**) it has much less influence on the strain of ACA joints

study of the effect of bump height showed that the failure mechanism of ACA flip-chip joints is totally different. In ACA joints, the bump and pad are usually made of metals that are much stiffer than adhesives. In other words, thermal mismatch stresses can hardly deform the bump and pad, and the shear strain is localized in the adhesive between the mating bump and pad (Fig. 5.13b). In this case, the joint reliability is governed by the shear strain in the adhesive and the influence of bump height is limited. Meanwhile, the stress in the *Z*-axis will be raised with bump height due to the increased adhesive volume. At elevated temperature, this stress can lift the chip and weaken the joint. So benefits from high bumps cannot be expected for ACA joints. Another practical problem associated with high bumps is that air bubbles are easily introduced during ACA bonding.

5.4.3 Degradation Due to Moisture Absorption

ACAs contain a much larger quantity of polymers. Therefore, polymer degradation due to moisture absorption becomes more significant in ACA joints. Water can degrade polymers through (1) depression of the glass transition temperature T_g and functioning as a plasticiser, (2) giving rise to swelling stresses, and (3) generating voids or promoting the catastrophic growth of voids already present. All three occurrences have been known to lead to mechanical degradation. Moisture absorption can also contribute to the disruption of conductivity in the path between mating electrodes. This may include, for example, changes in the polymer/filler dispersion state through the expansion of the polymer matrix and formation of defects such as cracks and delaminations.

The effects of moisture on an ACA film was studied with Fourier transform infra-red (FTIR) spectra that provide a vast reservoir of molecular information pertaining to the chemical groups present, as well as to the structure arrangement and bonding preferences of these groups. The adhesive was conditioned in two environments: 85°C/85%RH and 22°C/97%RH. After certain amount of time, samples were taken out of the chamber and FTIR spectra were collected.

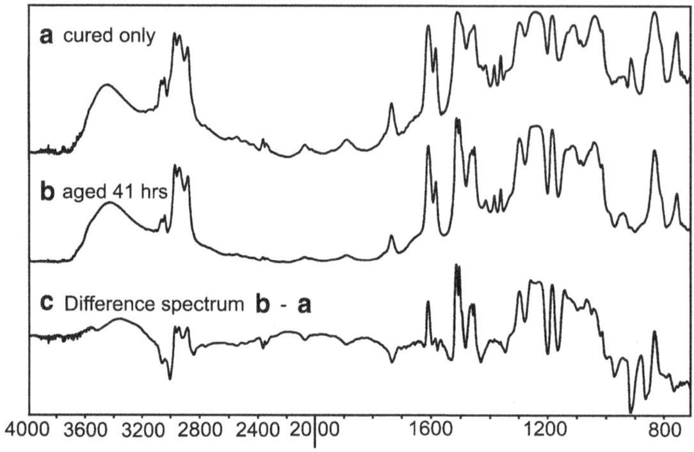

Fig. 5.14 FTIR spectra of an ACA film (**a**) after curing; (**b**) aged at 85°C/85%RH for 41 h, and (**c**) the difference spectrum (**b–a**)

Figure 5.14 shows the spectra of the adhesive (a) after curing, (b) after 41 h exposure to 85°C/85%RH, and (c) the difference spectra representing the changes due to the moisture exposure. As shown, the negative bands at 868, 916, 1,345, 3,005, and 3,058/cm indicate the further progress of the cure reaction. Moisture degradation is believed to occur by hydrolysis of the ester linkages, which creates two end groups: a hydroxyl and a carbonyl. Though it is hard to see any new emerging carbonyl groups in this figure, the band at 3,560/cm indicates the existence of free hydroxyls. With more time exposure, curing effect is not observed, but degradation becomes more apparent. The spectra collected from samples exposed to 22°C/97%RH showed that moisture absorption through hydrogen bonding, but neither further curing nor degradation is observed, implying that the dominant degradation is associated with heat.

5.4.4 Oxidation and Crack Growth

To correlate the electrical resistance shift as a function of humidity test time, a theoretical model has been developed. It takes into account both oxidation and cracking, two primary failure mechanisms of conductive adhesive joints and can thus explain the experimental observations quite well (Fig. 5.15).

Before exposure to the humid environment, the initial resistance through the joint is:

$$R_{init} = R_s + R_j + R_l, \tag{5.2}$$

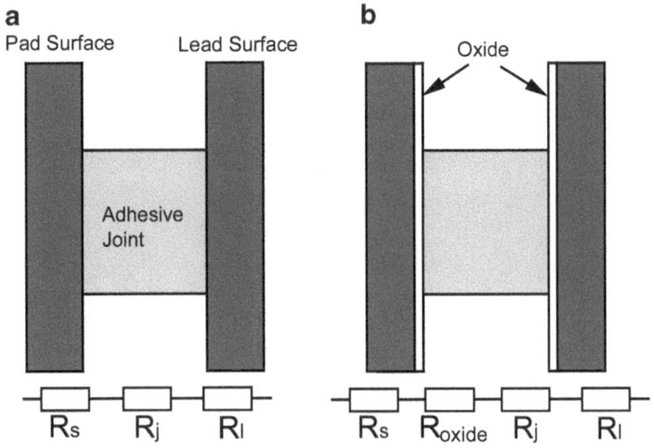

Fig. 5.15 Electrical conducting path through a conductive adhesive joint (**a**) before and (**b**) after humidity exposure

where R_s is the resistance through the substrate, R_j the resistance through the adhesive joint, and R_l the resistance through the component lead. After the humidity test, the joint resistance becomes:

$$R_{after} = R_s + R_j + R_l + R_{oxide} = R_{init} + R_{oxide}, \tag{5.3}$$

where R_{oxide} is the resistance through the oxide layer which can be expressed as:

$$R_{oxide} = \rho_{oxide} \frac{L}{A}, \tag{5.4}$$

where ρ_{oxide} is the volume resistivity of the oxide layer, L the oxide layer thickness, and A the contact area.

Since polymer structures normally contain a large amount of free volume, it is reasonable to assume that the diffusion of oxygen is much faster in polymers than in metal oxides. In other words, the oxygen diffusion through the oxide layer will control the oxide growth rate and consequently the increase of the resistance in the oxide layer.

Assume the following Einstein equation holds:

$$L = \sqrt{2D_{oxide}t}, \tag{5.5}$$

where D_{oxide} is the diffusion parameter of oxygen through the oxide layer and t is the time for the oxygen diffusion. Combining (5.3)–(5.5), one can obtain the relationship between the time and the resistance change:

$$\frac{R_{after}}{R_{init}} = 1 + \frac{L_e \rho_{oxide}}{AR_{init}} \sqrt{\frac{t}{t_e}}, \tag{5.6}$$

where L_e is the oxide layer thickness at the end of test, t the elapsed time, and t_e the total test time. Equation (5.6) can be used to calculate the relative electrical resistance change due to oxidation.

The crack normally occurs at the interface between the adhesive and the electrode and decreases the real contact area gradually. Here assume the contact area A can be expressed as:

$$A = A_0\left(1 - \frac{t}{t_e}\right), \tag{5.7}$$

where A_0 is the original contact area. Therefore, taking into account the crack growth, the electrical resistance change becomes:

$$\frac{R_{after}}{R_{init}} = 1 + \frac{L_e \rho_{oxide}}{A_0(1 - t/t_e)R_{init}} \sqrt{\frac{t}{t_e}}. \tag{5.8}$$

Figure 5.16 shows the calculated results with (5.6) and (5.8), using the parameters given in Table 5.1. The calculations show that if no crack is formed, the electrical resistance will increase gradually with test time, but no catastrophic failure will be expected. The effect of cracking is rather small at the beginning, but then becomes more and more significant with the increase of test time. If a complete crack forms by the end of the testing, the electrical resistance will go to infinity.

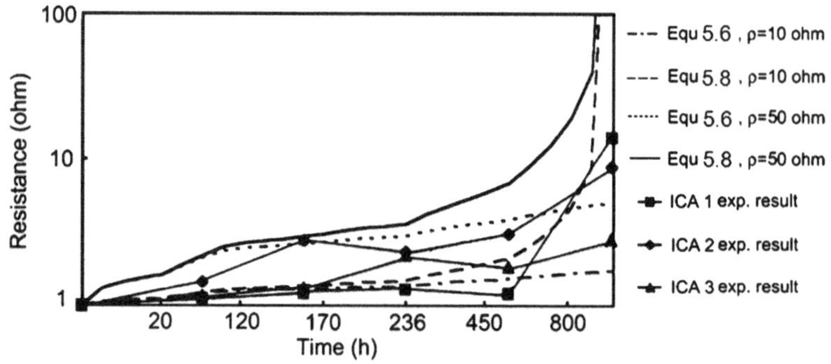

Fig. 5.16 Calculated and observed results of electrical resistance change as a function of humidity test time for ICA joints on copper surfaces

Table 5.1 Parameters used for calculation of the resistance evolution of an adhesive joint at 85°C/85%RH

Bonding surface	Oxide	Oxide (m)	A_0 (μm^2)	L_e (nm)	Dioxide (m^2/s)	R_{init} (−)
Copper	Cu_2O	10–50	1.1×10^{-6}	20	5×10^{-20}	0.2

For comparison, the experimental results obtained earlier are also given in Fig. 5.16. Before 500 test hours, (5.6) can predict the experimental observations quite well. However, the experimental results after 500 h cannot be explained by considering the oxidation of copper metal surface only, which means that fracture must have taken place during the humidity testing. In fact, cracks have already been observed after 158 h of exposure.

5.4.5 Probabilities of Open and Bridging

If the ACA contains insufficient particles, there is of course a certain probability that no particle exists in the joint and an open is resulted. On the other hand, bridging is possible due to there being too many particles in a too short spacing, causing short circuit between neighboring pads. Accurately estimating probabilities of open and bridging is important to explore the limiting pitch of ACA interconnection at which the open/short circuit probability becomes unacceptable (Fig. 5.17).

Mannan et al. proposed an analytical method to estimate the open probability. Assume that the number of particles on a pad obeys Poisson distribution:

$$P(n) = \frac{e^{-\mu}\mu^n}{n!}, \tag{5.9}$$

where $P(n)$ is the probability of finding n particles on a pad and μ is the average number of particles on a pad. If the volume fraction of particles f and the particle radius r are known, μ is given by:

$$\mu = \frac{3Af}{2\pi r^2}, \tag{5.10}$$

where A is the pad area. Thus the probability for an open ACA joint is:

$$P(0) = e^{-\mu} = e^{-3Af/2\pi r^2}. \tag{5.11}$$

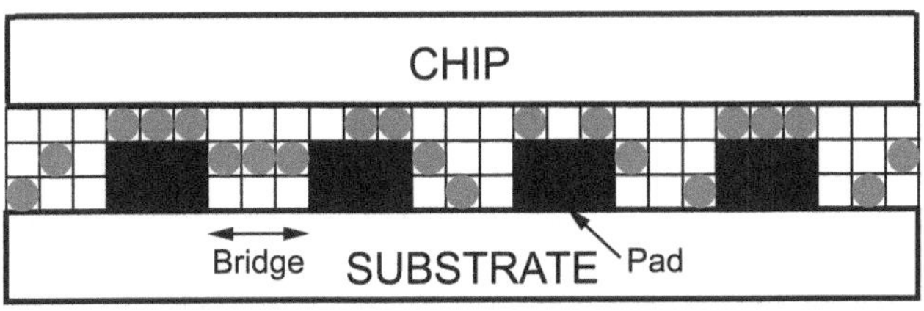

Fig. 5.17 Schematics of bridging in the ACA interconnection [20]

For a typical ACA with a volume fraction of particles ranging from 3 to 15 vol%, the open circuit probability on a 100 μm^2 pad varies from 10^{-13} to 10^{-3}, which is extremely small. However, in reality, there is always a crowding effect that must be taken into account. In this case, the particle distribution can be described using a binominal distribution model:

$$P(n) = C_n^N (1-s)^{N-n} s^n, \tag{5.12}$$

where N is the maximum number of particles that can be contained in the pad area A. C_n^N is the binominal coefficient and s is equal to f/f_m where f_m is the volume fraction corresponding to maximum packing. In the limit that $f \ll 1$, (5.11) and (5.12) give identical results for $P(0)$.

For a rough estimation of bridging, Mannan et al. proposed a simplified box model. As shown in Fig. 5.18, the volume between pads can be divided into cubic boxes with sides the same length as the particle diameter. If k boxes are filled out of a total of N, the volume fraction of particles is:

$$f = \frac{k4/3\pi r^3}{N(2r)^3}, \tag{5.13}$$

where r is the particle radius. Thus the probability for a single box being occupied is given by:

$$\frac{k}{N} = \frac{6f}{\pi}. \tag{5.14}$$

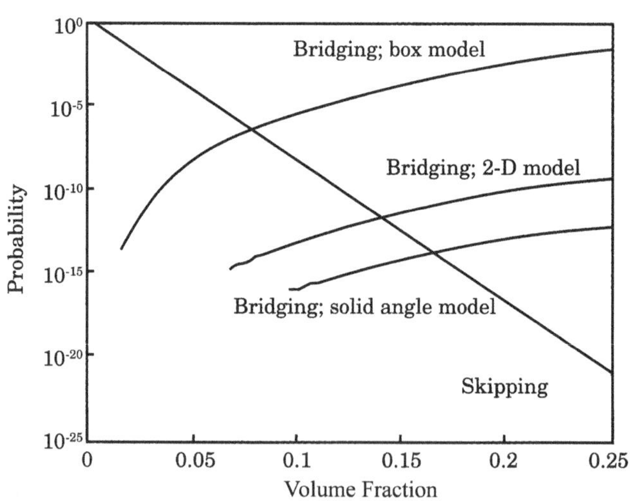

Fig. 5.18 Probability of particles bridging gap as a function of filler volume fraction [20]

Determined by the number of boxes that can be fitted onto the side of a single pad and by $(6f/\pi)^q$ where q is the lowest number of particles needed to bridge the pad spacing, the bridging probability is given by:

$$p = 1 - \left(1 - \left(\frac{6f}{\pi}\right)^{d/4r^2}\right)^{hl/4r^2}, \tag{5.15}$$

where h and d are the pad height and length, respectively, and l is the spacing between the pads.

This box model only gives an upper limit. Figure 5.18 shows the bridging probabilities derived from different models. It is clear that the lowest combined probability for bridging and skipping occurs in the volume fraction between 7 and 15%, depending on which model is used. This volume fraction range is also generally used for commercial ACA materials today.

5.4.6 ACA Flow During Bonding

As modeled by Mannan et al. [20], there are two types of adhesive flow during the ACA bonding (Fig. 5.19). Type 1 flow occurs around individual pads and bumps at the beginning of bonding, filling voids nearby. After voids are completely filled, type 2 flow becomes dominant, expelling the adhesive from under the chip to edges.

By solving the Navier–Stokes equations of Newtonian fluid, one can obtain the following equation that describes the pressure distribution under the chip in the cylindrical coordinate system:

$$P(r) = \frac{2F}{\pi R^2}\left(1 - \frac{r^2}{R^2}\right), \tag{5.16}$$

where R is half of the side length of the chip and F is the bonding force. In reality, the ACA resin probably behaves more like power law fluids:

$$\tau_{xy} = \eta_0\left(\frac{d\gamma}{dt}\right)^n, \tag{5.17}$$

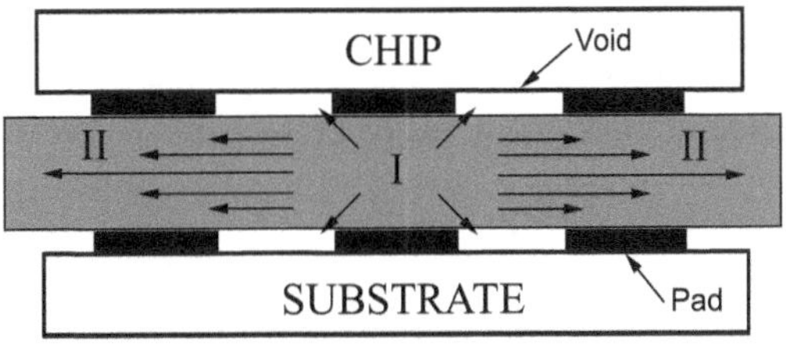

Fig. 5.19 The ACA flow during the bonding [20]

where η_0 is termed the consistency and n the power law index. For a Newtonian fluid, n equals 1 and η_0 becomes the viscosity of the fluid. As the chip is pressed down, the ACA is squeezed out between the chip and substrate. With power law fluids, the process time t_p for reducing the gap height from h_0 to h_1 is given by:

$$t_p = \frac{2n+1}{n+1}\left(\frac{2\pi\eta_0 R^{n+3}}{F(n+3)h_0^{n+1}}\right)^{1/n}\left(\left(\frac{h_0}{h_1}\right)^{(n+1)/n} - 1\right). \tag{5.18}$$

This process time is important to determine the suitable heating ramp for bonding. Too high bonding temperature may cause the adhesive solidify before particles are deformed completely, resulting in less reliable joints.

5.4.7 Electrical Conduction Development and Residual Stresses

As ACAs contain a small volume fraction of particles, there is no conduction in any direction before bonding. The electrical resistance starts to decrease as pressure increases due to enlarged contact area. Several research groups have reported the deformation effect on the electrical conduction development during the ACA assembly. The first publication is from Williams et al. [21] and the contact resistivity ρ of an ACA joint was estimated as:

$$\rho = \frac{A\rho_B\left(\sqrt{6\pi n\kappa/\sigma A} - (1/R_B)\right)}{4\pi n R_B}, \tag{5.19}$$

where ρ_B is the resistivity of the conductive particle, n is the number of contacts within the contact area A, κ is the shear yield stress of a conductive particle with a radius of R_B, and σ is the pressure applied to the joint.

With a combination of analytical method and FEM, Hu et al. [5] derived the relationship between the resistance and bonding pressure both for the rigid and deformable particle systems, as shown in Fig. 5.20. They also simulated the contact between the particle and electrode with FEM. As shown in Fig. 5.21, significant compressive stress is found to build up in the interface between the two contacts. This stress is believed to generate peel stress in the adhesive, which is probably the reason for catastrophic failure.

Fu et al. [6] considered the multiparticle case and found that the particle location in an ACA joint can affect its electric conductance. As shown in Fig. 5.22, a particle in the center of the joint contributes much more to the electrical performance than a particle close to the edge of the joint. This helps to explain why the measured resistance scatters greatly from one joint to another. Increasing the number of particles on the contact pad can improve the uniformity of the electric conduction.However, it also increases the constriction resistances due to fellow particles. So the total conductance does not increase in an additive manner.

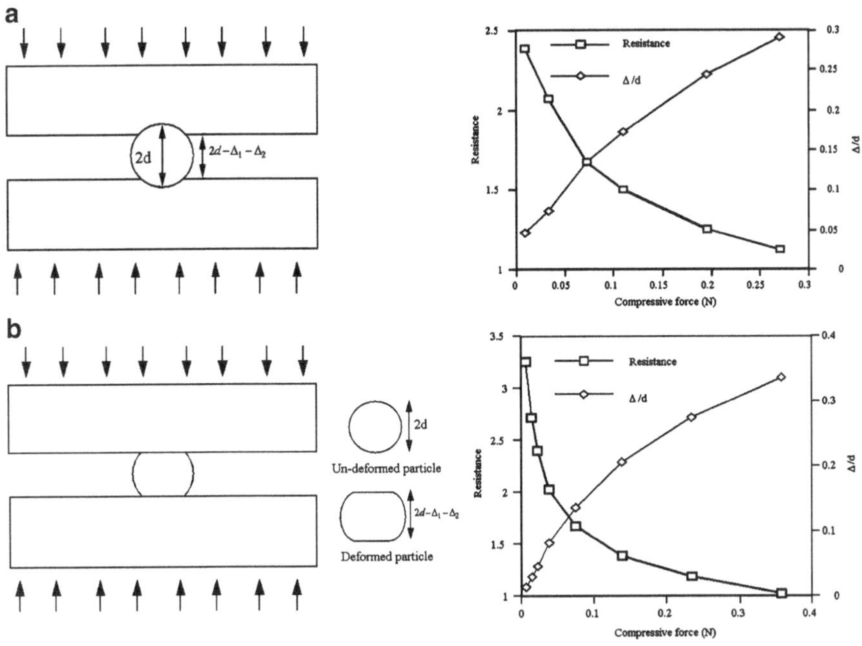

Fig. 5.20 Force–resistance–deformation relationships for (**a**) rigid particle system and (**b**) deformable particle system (courtesy of C.P. Yeh)

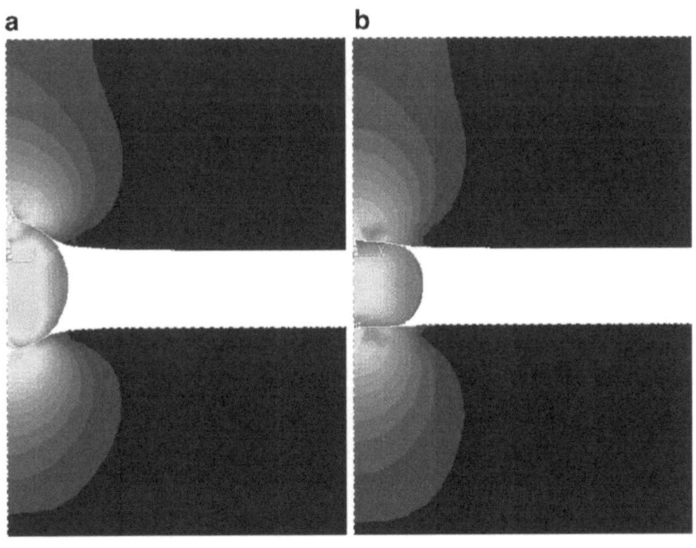

Fig. 5.21 Deformation distributions of (**a**) rigid particle system and (**b**) deformable particle system (courtesy of C.P. Yeh)

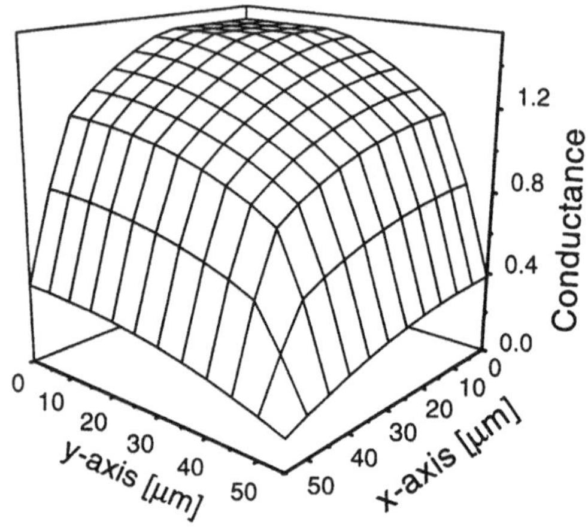

Fig. 5.22 Electric conductance of the particle as a function of its location away from the center of the ACA joint

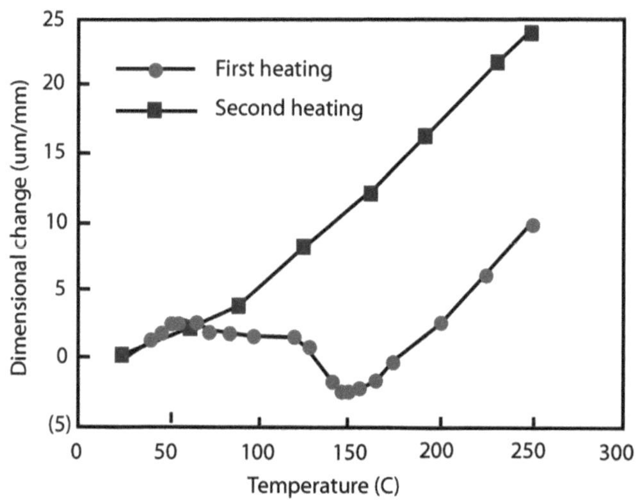

Fig. 5.23 Dimensional change of conductive adhesive

Exercises

5.1 Figure 5.23 shows the dimensional change during cure of a conductive adhesive. How can estimate the curing shrinkage from the curve. If the original sample is 60-μm long and the shrinkage is isotropic, what is the shrinkage value in this case?

5.2 What are the advantages of conductive adhesives comparing to traditional tin–lead solders?
5.3 For a thermal-setting conductive adhesive, the cure reaction can be expressed by Arrhenius equation as:

$$\frac{dx}{dt} = Z\exp\left(-\frac{E_a}{RT}\right)(1-x)^n.$$

The preexponential is 3.23×10^{14}/s, the activation energy is 106.4 kJ, R is gas constant as 8.314 J/mol/K. Assuming the curing reaction is second order, at least how long time should we hold it at 140°C before it is fully cured (99.99%)?
5.4 Why is Coffin–Manson relationship more suitable for solder joints and why is Morrow's law more suitable for conductive adhesive joints?
5.5 Why using higher bumps in the conductive adhesive joints cannot get a similar result in low cycle fatigue tests?

References

1. H. Botter, R. B. Van Der Plas and A. Arunjunai, "Factors that Influence the Electrical Contact Resistance of Isotropic Conductive Adhesive Joints During Climate Chamber Testing", International Journal of Microelectronic Packaging Materials and Technologies, 1, 1998, 177–185.
2. J. C. Jagt, "Reliability of Electrically Conductive Adhesive Joints for Surface Mount Applications: A Summary of the State of the Art", IEEE Transactions on Components Packaging and Manufacturing Technology Part A, 21, 1998, 215–225.
3. S. Y. Xu and D. A. Dillard, "Determining the Impact Resistance of Electrically Conductive Adhesives using a Falling Wedge Test", IEEE Transactions on Components and Packaging Technologies, 26, 2003, 554–562.
4. M. G. Perichaud et al., "Reliability Evaluation of Adhesive Bonded SMT Components in Industrial Applications", Microelectronics Reliability, 40, 2000, 1227–1234.
5. L. Li et al., "Reliability and Failure Mechanism of Isotropically Conductive Adhesives Joints", 45th Electronic Components and Technology Conference (ECTC), Proceedings, Las Vegas, NV, IEEE CPMT Society, 1995, 114–120.
6. S. Y. Xu, D. A. Dillard and J. G. Dillard, "Environmental Aging Effects on the Durability of Electrically Conductive Adhesive Joints", International Journal of Adhesion and Adhesives, 23, 2003, 235–250.
7. D. Lu, Q. K. Tong and C. P. Wong, "Mechanisms Underlying the Unstable Contact Resistance of Conductive Adhesives", IEEE Transactions on Electronics Packaging Manufacturing, 22, 1999, 228–232.
8. M. Yamashita and K. Suganuma, "Degradation Mechanism of Ag-Epoxy Conductive Adhesive/Sn-Pb Plating Interface by Heat Exposure", Journal of Electronic Materials, 31, 2002, 551–556.
9. R. L. Keusseyan, J. L. Dilday and B. S. Speck, "Electric Contact Phenomena in Conductive Adhesive Interconnections", International Journal of Microcircuits and Electronic Packaging, 17, 1994, 236–242.

10. R. S. Rörgren and J. Liu, "Reliability Assessment of Isotropically Conductive Adhesive Joints in Surface-Mount Applications", IEEE Transactions on Components Packaging and Manufacturing Technology Part B, 18, 1995, 305–312.
11. Z. M. Mo et al., "Electrical Characterization of Isotropic Conductive Adhesive under Mechanical Loading", Journal of Electronic Materials, 31, 2002, 916–920.
12. J. H. Constable et al., "Continuous Electrical Resistance Monitoring, Pull Strength, and Fatigue Life of Isotropically Conductive Adhesive Joints", IEEE Transactions on Components and Packaging Technologies, 22, 1999, 191–199.
13. H. K. Kim and F. G. Shi, "Electrical Reliability of Electrically Conductive Adhesive Joints: Dependence on Curing Condition and Current Density", Microelectronics Journal, 32, 2001, 315–321.
14. Y. Fu, J. Liu and M. Willander, "Conduction Modelling of a Conductive Adhesive with Bimodal Distribution of Conducting Element", International Journal of Adhesion and Adhesives, 19, 1999, 281–286.
15. L. Ye et al., "Effect of Ag Particle Size on Electrical Conductivity of Isotropically Conductive Adhesives", IEEE Transactions on Electronics Packaging Manufacturing, 22, 1999, 299–302.
16. Y. Fu and J. Liu, "Electron Conduction through Nano Particles in Electrically Conductive Adhesives", Micromaterials and Nanomaterials, 4, 2004, 104–109.
17. D. Klosterman, L. Li and J. E. Morris, "Materials Characterization, Conduction Development, and Curing Effects on Reliability of Isotropically Conductive Adhesives", IEEE Transactions on Components Packaging and Manufacturing Technology Part A, 21, 1998, 23–31.
18. A. J. Lovinger, "Development of Electrical-Conduction in Silver-Filled Epoxy Adhesives", Journal of Adhesion, 10, 1979, 1–15.
19. J. Liu et al., "Surface Characteristics, Reliability, and Failure Mechanisms of Tin/Lead, Copper, and Gold Metallizations", IEEE Transactions on Components, Packaging, and Manufacturing Technology Part A, 20, 1997, 21–30.
20. S. W. Kim, J. W. Yoon and S. B. Jung, "Interfacial Reactions and Shear Strengths Between Sn-Ag Based Pb-free Solder Balls and Au/EN/Cu Metallization", Journal of Electronic Materials, 33 (10), 2004, 1182–1189.
21. A. Sharif, Y. C. Chan, M. N. Islam and M. J. Rizvi, "Dissolution of Electroless Ni Metallization by Lead-Free Solder Alloys", Journal of Alloys and Compounds, 388(1), 2005, 75–82.

Chapter 6
Accelerated Testing

Abstract To perform reliability tests within a reasonable amount of time, accelerated tests are carried out in a laboratory environment in well-controlled conditions. The test condition has, therefore, to reproduce the real service conditions in an accelerated manner to achieve the same fracture mode.

Accelerated testing is conducted to determine the useful life of a certain component in the required product application. The main purpose of the accelerated test is to identify and quantify the failure and failure mechanisms that cause the component to fail.

Different accelerated tests are performed for each potential failure mechanism, as the stresses which produce failures are different for each mechanism.

There are many different failure mechanisms in microsystem products, and failures can be caused by thermomechanical, electrical, chemical, and/or environmental mechanisms.

This chapter will focus on accelerated testing of solder joints because of the prevalent use of solder in current practice. It describes both mechanical and thermal fatigue testing and the influence of different parameters on such tests, such as test frequency, stress/strain level, environmental conditions (temperature), ramp rate, and dwell time.

6.1 Fatigue Failure Analysis for Accelerated Testing

Failure mechanisms in microsystem products are many, and failures can be caused by either thermomechanical, electrical, chemical, and environmental mechanisms or a combination of the same. For a flip-chip PBGA, for example, typical failure modes can be underfill delamination, heat sink adhesive delamination, die cracking, substrate failure, PWB interconnection failure, and last but not least solder fatigue failure. The present chapter is concentrated on reliability aspects of solders and solder joints, and since thermomechanical fatigue is the main failure mechanism for solder joints, electrical, chemical, and environmental mechanisms are disregarded in the context of this chapter.

J. Liu et al., *Reliability of Microtechnology: Interconnects, Devices and Systems*,
DOI 10.1007/978-1-4419-5760-3_6,

The driving force for solder joint fatigue is the thermal mismatch between the various materials in a package, resulting in significant thermal stresses/strains. Besides residual stresses generated after assembly, solder joints are particularly subjected to severe shear strains, which are the major source for solder joint fatigue.

The fatigue process begins with the accumulation of damage at a localized region or regions due to the alternating load, which eventually leads to the formation of cracks and their subsequent propagation. When one of the cracks has grown to such an extent that the remaining cross-sectional area is insufficient to carry the applied load, a sudden fracture takes place. For macroscopically isotropic materials and during fatigue, Persistent Slip Bands (PSB) are major nucleation sites for cracks. Once cracks have initiated, they grow as a result of further cyclic deformation. Fatigue crack propagation generally occurs in two stages: stage I crack growth, which takes place along slip planes or planes of maximum shear and extends only a few grain diameters from the initiation siteand Stage II in which the crack follows a plane that is on an average perpendicular to the tensile axis. Due to high temperature or corrosive environments, cracks may also initiate at grain boundaries and propagate along the same [1].

Provided that a single failure mechanism is dominant, with a temperature dependent rate:

$$r = r_0 e^{-E_0/KT}, \tag{6.1}$$

where r_0, E_0 are characteristics of the failure mechanism in question (e.g., diffusion, corrosion, etc.), the times to failure t_1, t_2, at temperatures T_1, T_2, are related by the failure rate acceleration factor:

$$\frac{t_1}{t_2} = \exp\frac{E_0}{k}\left(\frac{1}{T_1} - \frac{1}{T_2}\right), \tag{6.2}$$

and $t_1/t_2 > 1$ for $T_2 > T_1$.

6.2 Thermal Fatigue

When executing thermal fatigue testing, the sample is subjected to temperature variations, and mechanical stresses arise in the solder joints due to the dissimilar CTEs of the different materials. There are different standards stating the test conditions that should be applied. During thermal fatigue, the temperature cycles are repeated with a certain time period until fracture occurs.

In service, solders are seldom subjected to regular continuous cycles. They normally experience dwell periods of several hours or days according to performance demands. These dwell periods at constant strain levels during which stress relaxation may occur introduce an additional factor influencing life span. Furthermore, decreasing the frequency normally produces a reduction in life span for solders [2–4].

At temperatures well below one half of the absolute melting point, however, frequency has little effect on the fatigue life of most materials [3].

Due to the temperature change, the material may exhibit quite different characteristics during the run of a single thermal cycle. This makes thermal fatigue a difficult phenomenon to analyze. In particular, the location of the dwell is critical since this controls the extent of time-dependent effects.

Deformation due to thermal stresses can be classified into thermoelastic, plastic, and creep. Elastic deformation is recoverable and is caused by changes in atomic spacing. Plastic deformation is permanent and is caused by dislocation motion. Creep is a time-dependent deformation which is caused by a diffusion process. The damage in the solder joints will be, therefore, a result of thermal-activated time-dependent mechanisms (creep), cyclic mechanisms (fatigue), and microstructural changes. These damage mechanisms are expected to interact with one another and to have different relative magnitudes. They also result in detectable fatigue damage quantities, such as elastic modulus degradation, plastic strain accumulation, and microstructure phase coarsening [5]. The elastic modulus of a solder material was observed to decrease as a function of number of cycles for thermal cycling tests performed on BGA packages. The elastic modulus degradation is considered to be directly related to macromaterial degradation under fatigue, and there is a relationship between the degradation in elastic modulus and plastic strain accumulation in the material, which is related to fatigue damage evolution [6].

Thermal fatigue cannot be predicted by using the standard Coffin–Manson relationship, which only takes into account the plastic strain range [6, 7] since it can lead to inaccurate damage quantification. Both the variation in temperature, which has a significant effect on the material properties and hysteresis strain energy dissipation, and the damage mechanisms under thermal loading are quite different from isothermal mechanical loading.

Furthermore, the load-drop criterion that is normally used in isothermal low-cycle fatigue (LCF) tests and is suitable to describe macrocrack propagation, cannot accurately describe the damage evolution of solder joints under thermal fatigue. The plastic strain accumulation in the solder joints during thermal cycling is a nonlinear process and the plastic strain range of just one or several cycles cannot appropriately reflect the physical mechanism of fatigue damage evolution. A modified Coffin–Manson equation has been presented [8], which takes into account the effect of temperature:

$$N_f = CF^m(\Delta T)^{-n} \exp\left(\frac{Q}{RT_{max}}\right), \tag{6.3}$$

where N_f is the thermal fatigue life, C a constant, F the frequency, ΔT the temperature range, Q the activation energy, R the gas constant, and T_{max} the maximum temperature.

To predict thermal fatigue behavior of solder joints, it is more accurate, however, to use the hysteresis energy-based damage which takes into account both strain and stress [7].

6.3 Effect of Different Test Factors on Thermal Fatigue Life

Thermal fatigue life is dependent on many different factors. It depends among other factors on the maximum (T_{max}) and minimum (T_{min}) temperature applied and the temperature range used ΔT. The larger the temperature differences, the higher the damage per cycle [9]. In general, the larger the maximum temperature T_{max}, temperature range ΔT, and dwell time and the faster the ramp rate, and the higher the stress level applied, the shorter the fatigue life [10]. The effect of heating rate on damage accumulation of Sn–Ag solder joints was investigated and found that a faster heating rate was more damaging compared with slower heating rate. The same results were obtained by Qi et al. [9]. Regarding hold time, increasing the hold time will decrease the fatigue life as a result of time-dependent creep.

Fatigue life definition has also an effect on thermal fatigue life. The Coffin–Manson cyclic strain-hardening exponent, α, was found to decrease when increasing the stress range drop parameter, Φ [$\Phi = 1-(\Delta\tau/\Delta\tau_{max})$]. It changed from 0.74 to 0.49 when changing the failure criterion from 10 to 50%. This variation of k as a function of failure definition reflects the difference in the rates of stress-range drop at different stages of cycling [11].

A summary of the effect of different factors on the thermal fatigue life of solder joints is shown in Table 6.1.

As pointed out before, creep is a phenomenon that also contributes to solder joint failure. The higher the temperature, the higher the contribution of creep. Creep strain is a result of thermally activated, time-dependent mechanisms. These mechanisms can obey a constitutive relation, as in (6.4):

$$\dot{\varepsilon} = C\left(\frac{1}{d}\right)^{a}\frac{1}{kT}\left(\frac{\sigma - \sigma_b}{E(T)}\right)^{n}\exp\left(-\frac{Q}{kT}\right), \tag{6.4}$$

Table 6.1 Effect of some factors on the thermal fatigue life of solder joints

Factor	Factor changes	Fatigue life changes	Comments
Frequency	Decrease	Decrease	At temperatures well below one half of the absolute melting point, frequency has little effect on the fatigue life of most materials
Hold time	Increase	Decrease	Hold time is much more destructive than ramp time (much lower strain rates operating during hold time)
T_{max}	Higher	Decrease	–
ΔT	Higher	Decrease	–
Heating/cooling rate(ramp rate)	Faster	Decrease	–
Failure definition	Decrease	Decrease	Higher strain hardening exponent α, at earlier stages of testing

where C is a constant, d refers to the grain size, a to the grain size sensitivity, σ is the applied stress with σ_b being the back stress, $E(T)$ the modulus as a function of temperature, and n is the stress exponent. The thermal activation of creep is characterized by an activation energy Q and k is the Boltzmann's constant. At high stresses, creep is controlled by dislocation movement. When dislocation entanglement and recovery reaches an impasse, where the rate of hardening is equal to the rate of recovery, a quasi-steady state is reached that obeys (6.4). This creep rate is then controlled by the rate at which edge dislocations can climb out of their slip planes. At lower stresses, creep is controlled by the motion of vacancies, from grain boundary to grain boundary. At the highest temperatures, vacancy motion happens by lattice diffusion and the creep is referred as Nabarro-Herring creep. At lower temperatures, vacancy motion happens through grain boundary diffusion and the creep mechanism is referred to as Coble creep. The damage that is stored during creep deformation can be of three types: creep cracks, void nucleation and growth, and microstructural degradation. Microstructural degradation is often the most serious damage in alloys that depend on the phase morphology for creep resistance. Stresses at high temperatures allow strengthening precipitates to coarsen and change shape, weakening the alloy.

For the steady state, creep follows the Garofalo–Arrhenius equation, expressed as:

$$\dot{\gamma} = C\left(\frac{G}{T}\right)\left[\sinh\left(\omega\frac{\tau}{G}\right)\right]^n \exp\left(-\frac{Q}{kT}\right), \tag{6.5}$$

where $\dot{\gamma}$ is the steady-state creep shear strain rate, t the time, C a material constant, G the temperature-dependent shear modulus, T the absolute temperature (K), ω defines the stress level at which the power law stress dependent breaks down, τ the shear stress, n the stress exponent, Q the activation energy for a specific diffusion mechanism (dislocation diffusion, solute diffusion, lattice self-diffusion, and grain boundary diffusion), and k the Boltzmann's constant (8.617×10^{-5} eV/K).

6.4 Isothermal Mechanical LCF

During isothermal mechanical fatigue testing, samples are cycled mechanically with a constant stress or strain amplitude, at a constant temperature. The testing executed with constant strain amplitude and where plastic strains are dominant is also called LCF.

A commonly used method to characterize LCF behavior of solder joints is the load/stress versus number of cycles. The pattern of load/stress reduction as a function of number of cycles can be described by a so-called load-drop parameter, defined in (6.6) as:

$$\Phi = 1 - \frac{\Delta F}{\Delta F_M}, \tag{6.6}$$

where ΔF is the load range at a certain load cycle number and ΔF_{M} is the maximum load range over the initial few cycles. The load-drop parameter curves can be divided into three different stages: the first called the rapid increase stage, the second the steady stage, and the third the acceleration stage. The steady stage is generally the dominating stage of the fatigue life and hence the slope of the load-drop parameter curve in the steady stage reflects the LCF life; the flatter the slope of the steady stage, the longer the fatigue life.

It is normally important to investigate, if the fatigue life is related to the applied plastic strain. The Coffin–Manson fatigue model is often used for the LCF analysis of solders. The Coffin–Manson relationship assumes that LCF failure is strictly a result of plastic deformation and the elastic strain has a negligible effect on the LCF life. The elastic strain range can also be included in the calculation, and the fatigue life is then defined in terms of both plastic and elastic strains. The relationship is given by:

$$\gamma_{\mathrm{t}} = \gamma_{\mathrm{e}} + \gamma_{\mathrm{p}} = \left(\frac{N_f}{C_B}\right)^{-1/\alpha} + \left(\frac{N_f}{C_{CM}}\right)^{-1/\alpha}, \tag{6.7}$$

In principle, both equations could be used to define fatigue life, N_{f}, for a given strain. For LCF applications, however, it is the correlation with the plastic strain that is used to predict fatigue life and since the elastic strain is generally very small in comparison to the plastic strain, which is the factor that really causes fatigue, this is normally ignored.

There is of course a third factor that is also important in the context of solder joint fatigue failure and that is creep. For solders, the cyclic creep effects are more pronounced at higher temperatures and slower test frequencies, decreasing the fatigue lives. Hence, the constants on the Coffin–Manson relationship are dependent on both test temperature and cyclic frequency. One disadvantage with the Coffin–Manson relation is that it only accounts for strain and not stress. For those reasons, another model that is increasingly being applied in the prediction of fatigue life of solder joints is the Morrow's energy density model. This model predicts fatigue life in terms of plastic strain energy density (W_p), and takes therefore into account both strain and stress:

$$N_f^m W_p = C, \tag{6.8}$$

where m is the fatigue exponent and C is the material ductility coefficient. The strain energy density is measured as the area of the hysteresis loops. The fatigue exponent and the material ductility coefficient are also dependent on test frequency and temperature.

The stress–strain history consists of the so-called hysteresis loops. The area of the hysteresis loop represents the energy dissipated in the material within one cycle. In the course of cyclic loading, materials can either harden or soften depending on their prior thermomechanical treatment. The primary hardening or softening period, which occurs quite rapidly in the early portion of fatigue life, is usually

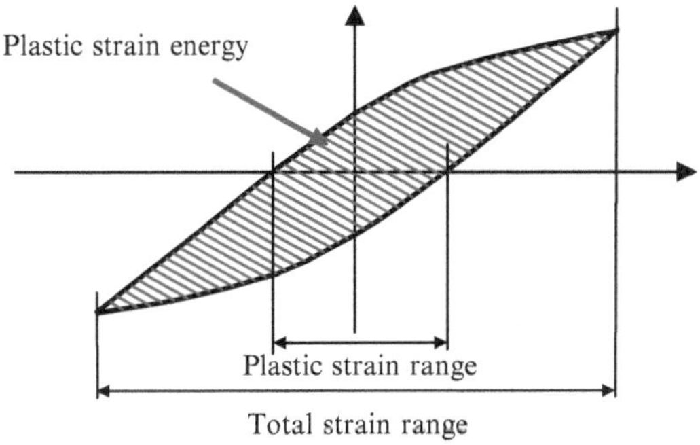

Fig. 6.1 Stress–strain hysteresis loop

followed by a steady-state cyclic deformation in which the stress–strain response remains constant. The hysteresis loop for a constant cyclic loading can be observed in Fig. 6.1.

The hysteresis loops provide very useful information for engineering evaluations of solder joint reliability. The width of the loop gives an estimate of the plastic strain range (intersection between the loop and the strain axis at zero stress). The total strain $\Delta\varepsilon$ is the sum of both the elastic and plastic strains.

6.4.1 Effect of Frequency

The effect of frequency on the isothermal mechanical fatigue life of most metals is dependent on the test temperature. For temperatures well below half of the absolute melting temperature, frequency has little effect on the fatigue life of most metals. Over this value, however, a reduction in frequency results in a decrease in fatigue life (N_f) for many metals, including solders. The reason for this behavior is that at high temperatures, creep failure, which is time dependent, plays a very important role in damage accumulation. Isothermal LCF tests performed on flip-chip solder joints showed that longer wave periods (slower frequency) leads to higher crack growth rates than shorter wave periods (higher frequency).

Some alloys, however, show frequency transition regimes, under or above which changes in frequency do not result in any appreciable fatigue life changes. For the Pb–3.5Sn, for example, the number of cycles to failure decreased steadily when cycling frequency was reduced below 10^{-2} Hz; however, no effect of frequency on N_f was detected at frequencies higher than 10^{-2} Hz. For this Pb-rich alloy, the effect of frequency was also found to be a function of strain range. The fatigue life of eutectic Sn–37Pb was also found to be frequency dependent over the test frequency range of 10^{-4} to 1 Hz. The decrease in fatigue life, however, was small when frequency decreased from 1 to 10^{-3} Hz, but became larger when the frequency was reduced further from 10^{-3} to 10^{-4} Hz. For the lead-free Sn–3.5Ag solder, the fatigue life also decreased as the frequency decreased from 1 to 10^{-3} Hz.

For other LCF tests performed with Sn–0.7Cu, increasing the frequency from 10^{-3} to 1 Hz significantly reduced the stress range and the plastic strain energy density. The fatigue life, tested at total strain ranges of 2.5 and 7.5%, at 398 K decreased linearly with decreasing frequency from 1 to 0.01 Hz.

The reduction in fatigue life with decreasing test frequency is attributed to the increasing exposure to creep and stress relaxation effects during fatigue testing. As the frequency decreases, the time for completing one cycle increases, which allows for longer exposure for creep and stress relaxation to develop and leads to a reduction in the stress range and hysteresis inelastic energy density.

To take into account the frequency during isothermal LCF tests, a frequency modified Coffin–Manson relationship can be used, which states:

$$\left[N_f v^{(k-1)}\right]^m \Delta\gamma_p = C, \tag{6.9}$$

where v is the frequency and k is the frequency exponent. Both ramp and hold time effects are considered because frequency is the inverse of the period, which is the sum of the ramp times and the hold times. The effect of frequency is determined by the magnitude of k. For $k = 1$, there is no dependence of the fatigue life with frequency variations and the frequency term is equal to 1 (for very low life, $N_f < 50$). When $k = 0$, the fatigue life is modified by $1/v$, and if the frequency is halved, the number of cycles to failure is also halved, which results in a constant time to failure [N/v is the time to failure which results in the development of a constant time to failure (for a given applied plastic strain range)].When the plastic strain range is constant, the fatigue life shows a linear relationship with frequency in a log–log plot, and where the slope of the curve is the value of $(1-v)$.

The frequency-modified Morrow model is the following:

$$\left[N_f v^{(h-1)}\right]^n W_p = A, \tag{6.10}$$

where v is the frequency and h is the frequency exponent. The frequency exponent k can be determined from the relationship between fatigue life and frequency. For a constant strain range, this relationship can be expressed as:

$$N_\mathrm{f} = bv^{1-k}, \tag{6.11}$$

where b is a constant and k is the frequency exponent.

6.4.2 *Effect of Dwell (Hold) Time*

In general, increasing the dwell time will decrease the fatigue life of solders. This is also a result of longer exposure to creep and stress relaxation. For lead-rich alloys, tested at room temperature and under a strain-controlled LCF, the dwell time was found to have a very high effect on fatigue life compared with other factors such as ramp rate, and increasing the dwell time decreased the fatigue life. Tensile hold times are more damaging compared with compressive hold times during high temperature fatigue of solders, and the fatigue life decreases when the tensile hold time is increased.

6.4.3 *Effect of Strain Range and Strain Rate*

Increasing the strain amplitude (strain range) results in a decrease in fatigue life [12]. The fatigue life of eutectic Sn–37Pb bulk alloy was found to decrease with increasing total strain range at a given temperature and frequency.

The effect of strain rate on the isothermal LCF of bulk Sn–37Pb was studied and the results showed a decrease in fatigue life with decreasing strain rate. They found, however, a transition regime at some intermediate strain rate and this relationship showed a typical S-shaped characteristic. The effect of strain rate on fatigue life became smaller with increased total strain range. The reason for this was found to be different failure mechanisms, where cavitation due to grain boundary sliding was the dominant failure mechanism in the low strain rate regime, while cavitation without grain boundary sliding was the dominant failure mechanism in the high strain rate regime. The transition strain rate was found to be ~10^{-3}–10^{-4} per second.

6.4.4 *Effect of Temperature*

In general, for all metals, an increase in temperature results in a decrease in isothermal fatigue life. The degree of fatigue life change depends, however, on the material and testing conditions. Above 0.6 T/T_m, the contribution of creep is expected to increase with increased temperature, which will result in shorter fatigue life. As the temperature increases the plastic strain range increases and the stress range decreases. It has been found, however, for a Pb-rich alloy, that the fatigue life dependency on temperature only follows approximately an Arrhenius equation between 25 and 80°C.

The eutectic Sn–37Pb alloy, tested as bulk material, was found to be temperature dependent over the range of test temperatures (−40 to 150°C). As the temperature increased, the fatigue life decreased linearly on a log–log plot.

6.4.5 *Effect of Failure Definition*

For isothermal LCF tests, changing the definition of failure will also affect the fatigue life. For LCF tests performed at room temperature, the fatigue life decreases as the load-drop failure definition decreases from 70 to 20% load drop. This conclusion is rather obvious; however, when comparing, and for that matter using, fatigue life data from different researchers, it is very important to know which failure definition was used since in addition to a decrease in fatigue life when decreasing the load-drop parameter, the slope of the plastic strain versus fatigue life plot will also change.

6.4.6 Effect of Other Factors

Many of the new lead-free solder alloys perform better in fatigue compared with the eutectic Sn–37Pb. Isothermal LCF tests performed at room temperature and at different loading angles showed that the fatigue life of Sn–3.5Ag–0.75Cu was longer for all loading conditions compared with the Sn–37Pb alloy.

The fatigue behavior of a solder alloy is affected by the addition of other elements. Under LCF tests of lap-shear samples, at room temperature and 0.1 Hz, the fatigue life of Sn–3.5Ag–xSb increases when increasing the amount of Sb from 1.73 to 10.05 wt% [13].

A summary of some factors and their effects on isothermal LCF life is depicted in Table 6.2.

The effect of creep on the fatigue life of the solder joints tested under isothermal LCF conditions (at room temperature) was not taken into consideration in the present work. By using relatively large strain range amplitudes, triangular wave shapes without any hold time and a relatively high frequency of 0.2 Hz decreases the effect of creep. For isothermal LCF tests performed at room temperature (25°C), the effect of creep can be disregarded when the testing frequency is higher than 10^{-3} Hz. It is known, however, that fatigue life is also dependent on test temperature, and the higher the temperature, the lower the fatigue life, which is a result of time-dependent creep.

Table 6.2 Overall effect of different factors on isothermal low cycle fatigue life

Factor	Factor changes	Fatigue life changes	Comments
Frequency	Decrease	Decrease	Dependent on test temperature: at $T < ½\, T_m$, frequency has little effect on fatigue life
Hold time	Increase	Decrease	Tensile hold time is more detrimental compared with compressive hold time. Hold time is more detrimental than ramp time
Strain range	Increase	Decrease	At a given temperature and frequency
Strain rate	Decrease	Decrease	–
Temperature	Increase	Decrease	–
Failure criteria	Decrease	Decrease	The slope of the plastic strain versus fatigue life plot will also change

Exercises

6.1 A FCOB assembly with a solder-bumped height of 0.1 mm and a distance from the neutral point of 0.3 mm is subjected to a cyclic temperature from −55 to +125°C. The coefficients of thermal expansion for silicon and FR-4 are 2.3 and 18 ppm/°C, respectively. For the simple case without underfill encapsulant, calculate the solder joint strain range, $\Delta\gamma$, and the fatigue life prediction model, N_f, as given by Engelmaier's model in question 4. If an undefill encapsulant is applied on the flip-chip assembly, discuss how the solder joint shear strain range can be estimated? Is finite-Element Analysis necessary? How does the estimated fatigue life compare with the case without encapsulant. The following parameters are given:

- $h = 0.1$ mm
- DNP = 3 mm
- $\Delta T = -55 + 125 = 180$°C
- $CTE_{silicon} = 2.3$ ppm/°C
- $CTE_{FR\text{-}4} = 18$ ppm/°C.

6.2 The homologous temperature for one metal is a rate between the temperature involved and its melting point. That is

$$T_h = \frac{T}{T_{melt}},$$

where the temperature is expressed in degrees absolute.
At homologous temperatures greater than 0.5, metals exhibit significant stress relaxation and creep. To describe the steady-state creep shear strain, Darveaux gave us the relationship as:

$$f(\tau) = C\left[\sinh\left(\omega\frac{\tau}{G}\right)\right]^n$$
$$g(t) = t$$
$$h(T) = \left(\frac{G}{T}\right)\exp\left(\frac{-Q}{kT}\right),$$

where τ is the shear stress, G the shear modulus, ω defines the stress level at which the power law stress dependence breaks down, k the Boltzmann constant, Q the activation energy, n the stress exponent, C a constant, and T the absolute temperature.
Calculate the homologous temperature for eutectic solder at room temperature (its melting point is 180°C). What is the Darveaux's constitutive law for eutectic solder? What is the steady-state creep shear strain rate?

6.3 Figure 6.2 below shows a CSP assembly. The chip was attached to the ceramic interposer (substrate) with gold bumps and underfill epoxy, and then soldered to the PCB. The chip size was 7 × 7 × 0.41 mm with 100 peripherally distributed gold bumps on a 0.25-mm pitch. The heights of the gold bumps were 0.025 mm as shown in Fig. 6.3 (not drawn to scale, and we only draw 20 instead of 100 gold bumps on the chip). The ceramic interposer dimensions were 7.45 × 7.45 × 0.25 mm. There were 100 arrayed eutectic solder bumps at the bottom of the ceramic substrate as shown in Fig. 6.4 (not drawn to scale, and we only draw 16 instead of 100 solder bumps on the ceramic substrate).

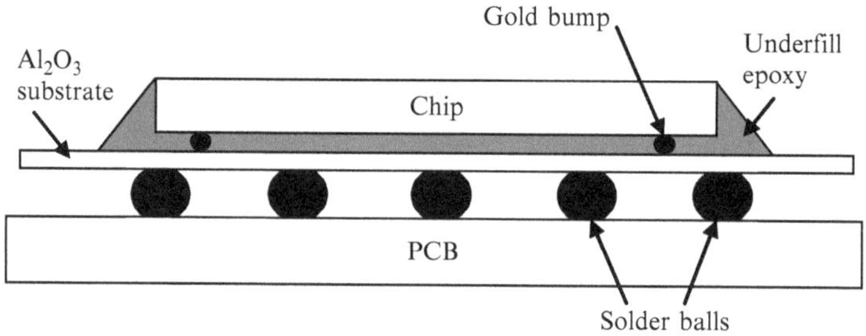

Fig. 6.2 Schematic cross section of a chip-scale package (CSP) assembly

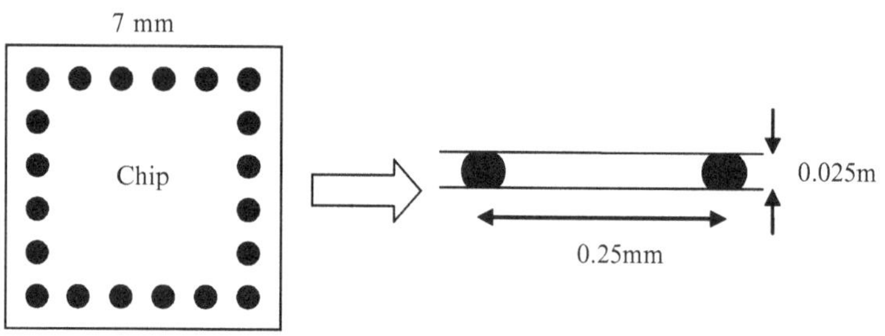

Fig. 6.3 Chip size and gold bump pitch for the CPS (not drawn to scale)

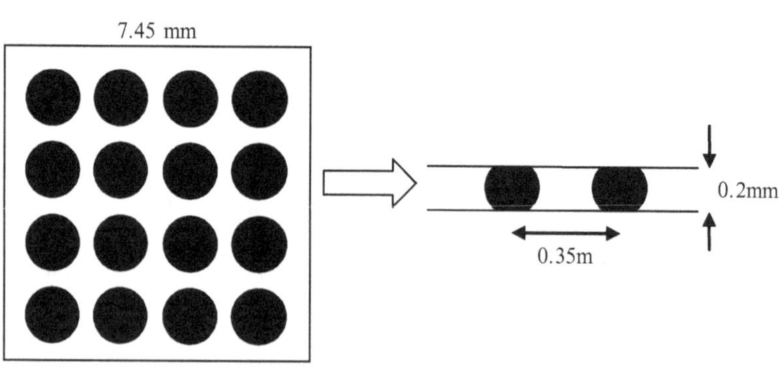

Fig. 6.4 Ceramic substrate and the eutectic solder balls on PCB (not drawn to scale)

After the CSP was assembled on the PCB, the solder ball height was about 0.2 mm.
A temperature loading is imposed on the CSP assembly from −40 to 125°C with 10-min ramp and 20-min hold. Te CTE mismatch between the chip, ceramic substrate, and the PCB brought the damage of the device (the typical CTE values are $\alpha_{chip} = 5$ ppm/K, $\alpha_{ceramic} = 10$ ppm/K, and $\alpha_{PCB} = 16$ ppm/K).

Question 1: What is the shear strain imposed on the gold bumps when it has no underfill?

Question 2: Consider the eutectic solder bump which is the farest away from the chip center (assuming the distance equals half of the ceramic substrate edge), what is the shear strain on it?

Question 3: Assuming both gold bumps and eutectic bumps in this study are obeyed the same Coffin–Manson equation with the exponential number −2, the lifetime for solder bumps is what times more than that of gold bumps? (this situation has been changed because of the existence of underfill)

Question 4: An interesting phenomenon is observed in experiments. For the eutectic bumps on PCB, the farther distance from the chip center, the easier they will be damaged. Can you explain it?

Given:

Chip size: 7 × 7 × 0.41 mm with 100 gold bumps on a 0.25-mm pitch Height of the gold
Bumps: 0.025 mm
Ceramic substrate:7.45 × 7.45 × 0.25 mm, 100 solder bumps (0.2 mm high after reflow)

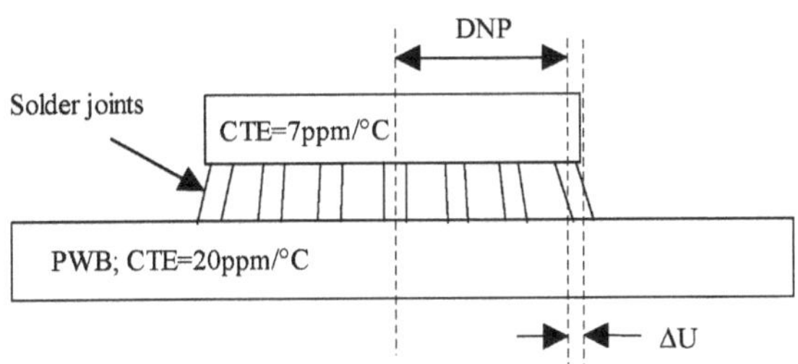

DNP (distance to neutral point)=L=11.4mm

Temperature cycle: −40 to 125°C with 10-min ramp and 20-min hold.

6.4 Ceramic ball grid array (CBGA) package.

Thermal cycling ⇒ different expansion of the different parts ⇒ the relative displacement ΔU of a solder joint is calculated from the difference between the top and the bottom surfaces of the solder joint.
When the temperature raises by 100°C, what is the relative displacement in the right end solder joint? Estimate the maximum shear strain range ($\Delta\gamma$) in the

Table 6.3 Life time and mean plastic shear strain range for type A and type B

	Type A	Type B
N_f	87	2,250
$\Delta\gamma$	0.0866	0.0101

solder joint of a perimeter PBGA package assembled onto an FR-4 PWB subjected to a temperature range of 0–100°C. The package has a DNP = 17 mm to the outermost solder joint and the solder height is 0.5 mm. The CTE of the BT (Bismaleimide Triazine) substrate is 15 ppm/°C and for the FR-4 PCB the CTE is 18 ppm/°C. The effective CTE of the mold compound and silicon die may be assumed to be the same as the BT substrate.

6.5 The solder joint fatigue life for the perimeter PBGA given in question 3 can be assessed using Engelmaier's Model for solder joint fatigue prediction. Two thermal cycling profiles are to be evaluated. The first temperature profile is from +25 to +125°C with a cycle time of 40 min. The second temperature profile is from −20 to +80°C with a cycle time of 24 min. Which temperature profile is more damaging in fatigue life?

6.6 The Coffin–Manson relation is based on that the solder joint failure is dependent on the accumulation of the plastic strain damage. It has been widely used to predict the thermal fatigue life of solder joints.

$$N_f = C(\Delta\gamma)^{\beta},$$

where N_f is the number of the cycles of failure, $\Delta\gamma$ is the plastic shear strain range, C and β are the material constants.There are two types of flip-chip electronic package (type A and type B) with the same material while different geometry. Table 6.3 gives the lifetime N_f measured by accelerated test and the plastic shear strain range $\Delta\gamma$ calaulated by FEM simulation.

Please calculate the empirical parameters C and β in the Coffin–Manson equation for this kind of flip-chip package using the data from Table 6.3.

References

1. F. Ellyin (Ed.), "Fatigue damage crack growth and life prediction", Chapman & Hall, London, 1997.
2. H. D. Solomon, "Fatigue of 60/40 solder", IEEE Transactions on Components, Hybrids and Manufacturing Technology, CHMT-9(4), 1986, 423–432.
3. S. Vaynman, "Fatigue life prediction of solder material: Effect of ramp time, hold time and temperature", Proceedings of the 40th Electronic Components and Technology Conference, 1, 1990, 505–509.
4. X. Q. Shi, H. L. J. Pang, W. Zhou, Z. P. Wang, "Low cycle fatigue analysis of temperature and frequency effects in eutectic solder alloy", International Journal of Fatigue, 22, 2000, 217–228.

5. C. Basaran, H. Tang, S. Nie, "Experimental damage mechanics of microelectronic solder joints under fatigue loading", Electronic Components and Technology Conference, 1, 2005, 874–881.
6. H. Tang, C. Basaran, "Experimental characterization of material degradation under fatigue loading", IEEE, The Eighth Intersociety Conference on Thermal and Thermomechanical Phenomena in Electronics, ITHERM, 2002, 896–902.
7. V. Sarihan, "Energy based methodology for damage and life prediction of solder joints under thermal cycling", IEEE Transactions on Components, Packaging and Manufacturing Technology-Part B, 17(4), 1994, 626–631.
8. I. Shohji, H. Mori, Y. Orii, "Solder joint reliability evaluation of chip scale package using a modified Coffin-Manson equation", Microelectronics Reliability, 44, 2004, 269–274.
9. Y. Qi, R. Lam, H. R. Ghorbani, P. Snugovsky, J. K. Spelt, "Temperature profile effects in accelerated thermal cycling of SnPb and Pb-free solder joints", Microelectronics Reliability, 46, 2006, 574–588.
10. H. Cui, "Accelerated temperature cycle test and Coffin-Manson model for electronic packaging", IEEE RAMS, 2005, 556–560.
11. X. W. Liu, W. J. Plumbridge, "Thermomechanical fatigue of Sn-37 wt% Pb model solder joints", Materials Science and Engineering A, 362, 2003, 309–321.
12. S. Vaynman, M. E. Fine, "Effects of strain range, ramp time, hold time and temperature on isothermal fatigue life of tin-lead solder alloys", in Solder joint reliability – theory and applications, edited by J. H. Lau, Van Nostrand Reinhold, New York, 1991.
13. H. T. Lee, H. S. Lin, C. S. Lee, P. W. Chen, "Reliability of Sn-Ag-Sb lead-free solder alloys", Materials Science and Engineering A, 407, 2005, 36–44.

Chapter 7
Reliability Design for Manufacturability

Abstract When looking at the manufacturing of electronic products and the use of lead-free solders, there are some issues that have to be considered. Issues such as alloy selection and paste handling, component type and component finishes, fluxes, and manufacturing process parameters, such as temperature, time, and atmosphere, have to be reevaluated when changing to lead-free solder manufacturing.

Lead-free solders present different physical properties compared with the conventional tin–lead solders. The most accepted lead-free alternatives present, for example, higher melting temperatures compared with the typically used Sn–Pb eutectic solder, which can affect both the manufacturability and reliability of lead-free electronics. Smaller process windows, damage to temperature sensitive components, and board warpage are only some examples of the problems that can occur while soldering with lead-free solders.

This chapter gives a short introduction to such issues, from the effect of higher processing temperature, failures resulting thereby, other defects connected to the fact that lead-free solders present different physical properties, inspection issues, repair, and rework of lead-free products. Other issues, such as lead contamination and tin whiskers are also shortly presented and discussed.

7.1 Lead-Free Soldering

7.1.1 Higher Process Temperature

Typical reflow soldering temperatures for Sn–Pb alloys have a peak temperature of ~220°C. Using lead-free solders results in higher process temperatures (245–255°C), which might take a significant toll on materials, components, and in some cases on the reflow equipment used.

At the component level, the higher reflow temperature can affect the die attach epoxy, mold compound, and substrate warpage and coplanarity. Components such as electrolytic capacitors and relays are very susceptible to temperature damage. The popcorn effect is another failure mechanism that is also greatly influenced by temperature. From the board perspective, the viability of the board material under

J. Liu et al., *Reliability of Microtechnology: Interconnects, Devices and Systems*,
DOI 10.1007/978-1-4419-5760-3_7,

the influence of high temperature can also be a concern. Standard PCB materials such as the glass/epoxy composite flame-retardant class 4 (FR-4) can be heated up to between 250 and 270°C. The temperature resistance of this material should be sufficient for most lead-free alloys, however, using some alloys that have higher melting points might result in exceeding reflow temperatures of 270°C. In this case, alternative PCB materials must be used. Examples of such materials are FR-5 with a T_g of ~180°C and a 1.5 times cost increase, and glass/BT polyimide with a T_g of ~250°C and a cost of 3.5–5.5 times the cost of FR-4 [1].

Fluxes have also to be considered, and they have to be stable at the higher temperatures and cannot cause shorts, contamination or corrosion. Compatibility between soldering temperature and chemical and physical properties is the key.

Other failure modes related to an increase in reflow temperature are the following:

- Corrosion of Al in the semiconductors caused by the popcorn effect during soldering. The higher reflow temperature increases the risk for delamination between interfaces in the plastic package. The moisture fills the voids and in combination with ionic contamination and a potential between conductors, corrosion of the aluminum will occur.
- Increased thickness of intermetallic compounds between Sn–Cu and Sn–Ni.
- Cracks and delamination in IC packages.
- Oxidation of boards, pads, and leads and degradation of certain laminate coatings.
- Electromigration and short-circuiting in PCBs.

Furthermore, higher melting temperature results in the margin between the minimum temperature for reliable reflow and the maximum temperature for materials' safety (component tolerance) shrink. The process window that exists for a trouble-free soldering becomes narrower and the need for profiling increases; see Fig. 7.1.

A direct consequence of higher melting points of lead-free alloys on the wave soldering process will be a higher pot temperature ranging between 255 and 270°C. Most modern wave soldering machines can provide the necessary heat (preheat and wave) for lead-free soldering, so that will not be a concern. However, due to higher melting temperatures, together with different chemical reactions that will occur in the pot, it will normally be necessary to change material for the pot, nozzles, impellers, and other parts that are manufactured of stainless steel and that are in

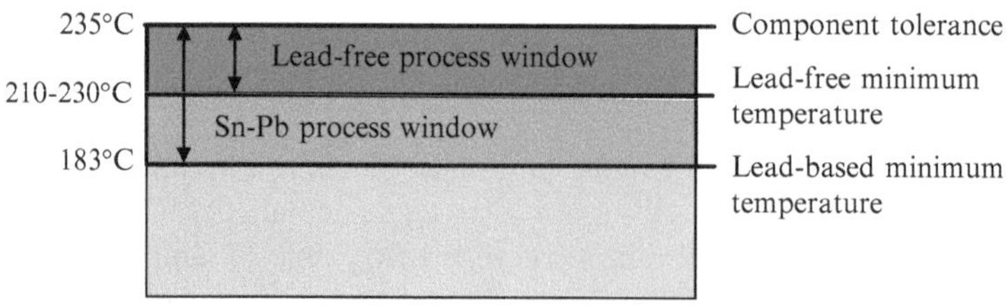

Fig. 7.1 Effect of lead-free reflow temperature on process window

contact with the molten solder. The reason for this is that when using high tin-content alloys at high temperatures, the tin will become corrosive to the stainless steel and this reaction will result in steel particles dissolving into the bath. Without protection such stainless steel pots and other parts will degrade within 1–2 years using lead-free solders. An alternative material to stainless steel is cast iron. Another possible alternative is to surface treat stainless steel pots to protect them from corrosion.

The higher pot temperature will also require a higher preheat temperature, this to thermally precondition the boards prior to their contact with the higher pot temperatures. Higher preheat and pot temperature, however, will cause the flux material to evaporate more easily prior to the wave. Consequently, new fluxes have to be designed to be able to handle these higher temperatures.

Another negative consequence of using lead-free solders in wave soldering is more dross formation. It is expected that, for some applications, N_2 atmosphere may be required although a fully inert machine seems not to be required. An additional problem is that solder baths are prone to lead contamination, especially when substrates and components are Sn/Pb coated. There are already lead-free board and component lead finishes available. Many of these, such as OSPs and Au/Ni, have been available for years. It is therefore imperative to have a 100% lead-free product to avoid such contamination problems.

7.2 Other Issues

Lead-free solders display higher surface tensions compared with Sn–Pb alloys, which results in increased wetting angles and is synonymous with less spread and worse wettability. Higher surface tension also results in higher voiding in lead-free joints, since it is more difficult for voids to escape; see Fig. 7.2a–f.

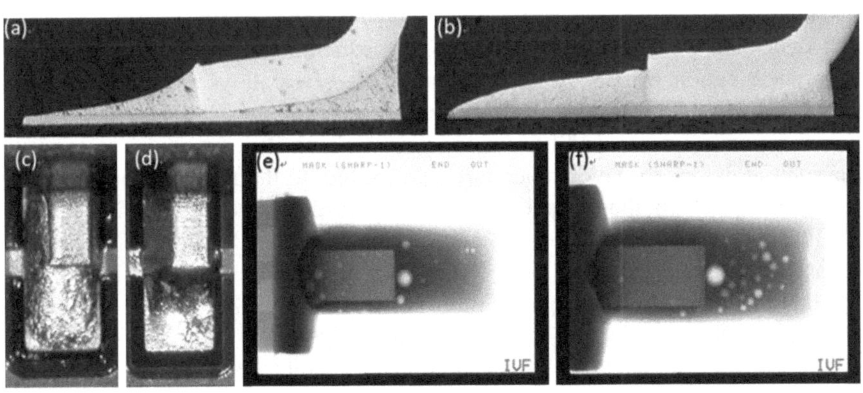

Fig. 7.2 (**a**) Cross-sectional view of SOP solder joint with Sn–37Pb alloy; (**b**) Cross-sectional view SOP joint with SAC solder; (**c**) Top view of joint depicted in (**b**); (**d**) Top view of solder joint depicted in (**a**); (**e**) X-ray image of SOP solder joint made of Sn–37Pb solder alloy; and (**f**) X-ray of SOP solder joint made of SAC solder alloy

Table 7.1 Surface tension and density for solder alloys [2–6]

Alloy	Surface tension γ (mN/mm) in air	Wetting angle	Density (g/cm^3)
Sn	550	–	7.37
Sn–37Pb	318[a], 470	14[b] (340°C)	8.40
Bi–42Sn	319[a], 350	43 ± 8 (195°C)	8.74
Sn–9Zn	518[a], 490	59[b] (250°C), 58[c]	7.27, 5.3
Sn–3.5Ag	431[a], 580	38[b] (250°C), 45[c]	7.39, 7.48
Sn–0.7Cu	491[a], 460	–	7.29, 7.15, 7.5
Sn–4Ag–0.5Cu	480	–	7.5
Sn–5Sb	468[a]	–	7.25

[a]At T_{liq} + 50°C
[b]K_{ester} Rosin flux no. 197
[c]20% Rosin in isopropyl alcohol, 250°C

Table 7.1 shows the surface tensions for some solder alloys, measured in air. The surface tension will vary from its ideal value in practical situations, depending on metallurgy, alloy purity, and atmosphere [2]. The surface tension and density are also dependent on the measurement temperature and decrease with increasing temperature [3].

The higher values of surface tension of lead-free solders might also result in a required higher accuracy in the placement of very fine pitch components, since self-alignment is not as effective as for Sn–Pb solder alloys.

7.2.1 Lead Contamination

The issue of lead contamination has not been acknowledged in the past. The logic behind this fact was that Sn and Pb are soluble in a lead-free system. It was forgotten, however, that the IMCs in lead-free systems are not soluble and will precipitate at lead boundaries [7]. In some cases, and as shown in Table 7.2, the presence of Pb in lead-free solders is likely to produce a phase, which melts about 40°C below the same combination without Pb [8, 9]. Certain lead-free alloys are more sensitive for Pb contamination than others. Especially sensitive are Bi-containing alloys. If these come in contact with base materials containing Pb, new phases are formed that melt at extremely low temperatures. According to the ternary phase diagram of Sn–Pb–Bi system, there is a ternary eutectic reaction at 96°C [L $\rightarrow$ X + (Sn) + (Bi)], where X is a metastable phase containing Bi, Pb, and Sn. At 78°C, X has a ternary eutectoid reaction [L $\rightarrow$ β + (Sn) + (Bi)], where β is Pb_3Bi IMC. Such low-melting phases have a negative influence on the reliability of the solder joint, especially concerning thermal fatigue at higher temperatures [10].

The wetting properties, including melting temperature and shear strength for Sn–2.5Ag–0.8Cu–0.5Sb (CASTIN) and for Sn–3.33Ag–4.83Bi, were studied as a function of Sn–37Pb contamination (0 < wt% < 10). Both alloys displayed lower melting temperature with higher contamination, and the Sn–Ag–Bi alloy was affected more by the Pb contamination than the CASTIN solder [11].

Table 7.2 Influence of Pb on some lead-free binary systems: the creation of low melting phases

Binary system	Lowest melting temperature in the binary system [°C]	Lowest melting temperature with Pb added to the binary system [°C]
Sn–Ag	221	178
Sn–Bi	139	96
Sn–Zn	199	183
Sn–Cu	227	183

Other researchers have concluded, however, that the Sn–Bi–Pb ternary phase (MP = 96°C) does not form measurable amounts in solder-only samples until a nominal proportion of 10.5% Bi and then disappears upon subsequent reflow cycles [12].

The effect of Pb contamination on the microstructure and mechanical properties of Sn-3.5Ag has also been investigated. After contamination with Pb, the microstructure of the Sn-3.5Ag alloy showed a third darker-colored phase that appeared at the grain boundaries of the bulk solder after reflow, proved to be a Pb-rich phase by means of EDX analysis. According to the shear strength tests performed, the Pb contamination did not have any influence on the same when the testing was performed at room temperature. However, when the shear testing was executed at 125°C, the shear strength of the solder joints without any Pb contamination was about 15% higher than that of those contaminated with Pb [13].

Other authors have shown that the reliability of lead-free solder joints is also considerably reduced by Pb contamination. However, the mechanism by which that took place did not involve low-melting-point phases. Instead, cracking was initiated at room temperature under stress at the Pb–Sn grain interface and then propagated along the Sn grain boundaries [14].

Despite the reason of reliability reduction, researchers seem to agree upon the negative influence of Pb contamination on the integrity of lead-free solders.

7.2.2 *Tin Whiskers*

A tin whisker is a spontaneous columnar or cylindrical filament (elongated single tin crystals), ranging normally from 6 nm to 6 μm in diameter and up to several millimeter long; see Fig. 7.3a. Tin whiskers are of great concern when using tin-plated or pure tin component finishes. Since the majority of all Pb-free solders contain very large amounts of Sn, they also show higher propensity for tin whiskers growth. The presence of tin whiskers, which can develop aspect ratios >1,000 (length/diameter), and are very brittle, can lead to shorting, and thereby threaten the reliability of an electronic device [15].

There are different theories regarding the growth of whiskers. A metallurgical theory explains the whisker growth as a result of the Sn crystal structure being

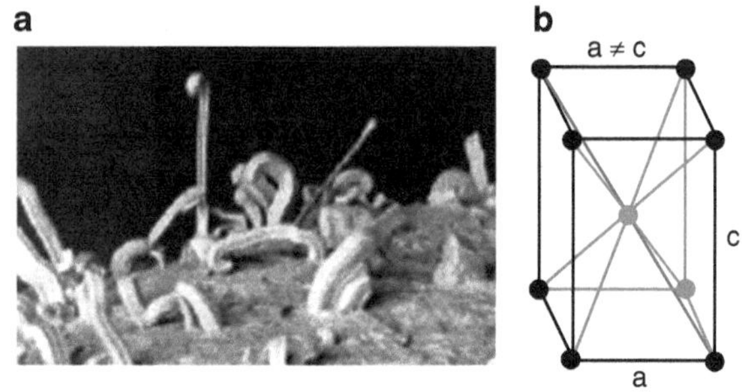

Fig. 7.3 (**a**) Typical Sn whiskers growing on a plated surface [15]; (**b**) bct crystal structure

anisotropic and having preferred slip planes. The white β-tin, which constitutes the tin whisker, has a body-centered tetragonal (bct) crystal structure and therefore is anisotropic; see Fig. 7.3b.

Another model that explains the growth of whiskers is a mechanical stress model, which states that internal compressive stresses are a driving force for the growth of tin whiskers. Critical precursors that increase the propensity of tin whisker formation are: compressive stresses in the plated coating, intermetallic formation between the tin and other metals, external mechanical stresses applied to the tin and to a certain degree, the grain structure of the plated tin. There are different ways of avoiding whiskers; avoid pure tin, and especially bright tin and use matte Sn which is less prone to whisker formation. Reflow of the tin plating to refuse/recrystallize and stress-relieve the deposit and using barrier metals to encapsulate any whiskers which were formed since the completion of the plating has also proved effective.

7.3 Inspection

Lead-free solder joints are duller and have a more grainy appearance than tin–lead solder joints (Fig. 7.4a) resulting in a need for operator training and inspection-machinery reprogramming. Workers trained for postsolder inspection are used to bright, shiny, and smooth joints! Duller joints have been the marker for poor joint quality when using conventional solder! This implies that a change in acceptance on how an acceptable solder joint should look like has to take place.

On SMT with leaded devices, the foot of the lead becomes more visible when using lead-free compared with tin–lead solders, due to a somewhat poorer wettability of lead-free solders; see Fig. 7.4b. Voiding is another issue that is accentuated in lead-free solders. Array packages, for example, tend to exhibit more solder-ball voiding when using lead-free solders compared with conventional Sn–Pb solder; see Fig. 7.4c.

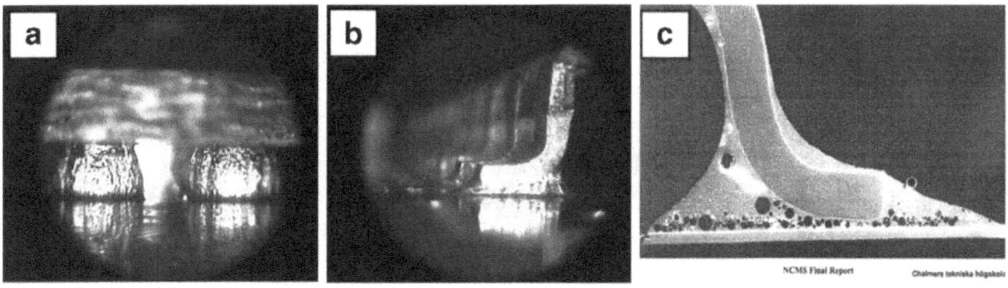

Fig. 7.4 (**a**) BGA lead-free solder joints exhibiting a duller appearance; (**b**) SOP joint; (**c**) Lead-free solder joint with voids

New workmanship standards have to be developed, and inspection equipment has to be reprogrammed for proper and accurate inspection of lead-free joints. Inspection workers have to be retrained in how to inspect these new joints. This is a one-time measure that hopefully will not affect production yields.

7.4 Repair and Rework

Since lead-free solders do not wet as well as tin/lead, operators have to be retrained for lead-free rework. An important aspect that has to be taken into consideration is the fact that different lead-free solders should not be mixed on the same joint. All rework should use the same lead-free solder alloy as originally used on the solder joint. A good rule of thumb is that mixed alloys can compromise reliability.

The removal and replacement of components is not expected to be a problem. Again, the only difference is expected to be higher iron tip temperature, which is not expected to affect the repair and rework step. It is, however, important to ensure that the desoldering and soldering stations are suitable for lead-free processes. They should be able to reach the necessary temperature for lead-free soldering. As a consequence of higher soldering temperature, there is a negative effect on the tip life of soldering irons. It has been reported that tip lives are shortened by enhanced erosion.

Exercises

7.1 Why are tin whiskers so deleterious?
7.2 What are the acceptance criteria for tin whiskers? What is the method of identifying tin whiskers?
7.3 What is being done to mitigate whisker growth?

7.4 Is nitrogen required in lead-free wave soldering?
7.5 Why are lead-free solders more susceptible for voiding compared with Sn–Pb solders? What can you do to reduce voiding?
7.6 Imagine you have to visually inspect two different boards; one soldered with tin–lead solder and one soldered with lead-free solder. How can you see which board is the lead-free one?
7.7 Why are Bi-containing alloys so sensitive for lead contamination?
7.8 What should you take into consideration when purchasing components to be soldered using a lead-free solder?
7.9 Explain the phenomenon of pop corning. How can you avoid it?

References

1. A. Z. Miric, A. Grusd, "Lead-free alloys," Soldering and Surface Mount Technology, 10(1), 1998, 19–25.
2. J. Glazer, "Metallurgy of low temperature Pb-free solders for electronic assembly," International Materials Reviews, 40(2), 1995, 65–93.
3. I. Kaban, S. Mhiaoui, W. Hoyer, J. G. Gasser, "Surface tension and density of binary lead and lead-free Sn-based solders," Journal of Physics: Condensed Matter, 17, 2005, 7867–7873.
4. M. Abteu, G. Selvaduray, "Lead-free solders in microelectronics," Materials Science and Engineering, 27, 2000, 95–141.
5. Z. Moser, W. Gasior, J. Pstrus, "Surface tension measurements of the Bi–Sn and Sn–Bi–Ag liquid alloys," Journal of Electronic Materials, 30(9), 2001, 1104–1111.
6. G. Poupon, "Remplacement du plomb dans les alliages de soudure en microélectronique," http://www.sansplomb.org/doc_presentation/pbissue.pdf, 2007.
7. "Wolfson microelectronics," http://www.wolfsonmicro.com
8. L. G. Klang, "Blyfria lod för mjuklödning-en litteraturstudie," Institutet för Metallforskning, research report, 1994.
9. W. B. Hampshire, "Some problems in switching to lead-free solders," Soldering and Surface Mount Technology, 9(1), 1997, 11–12.
10. "STATS, lead-free/halide-free update," http://www.stts.com/stts/docs/products/lead-free.pdf
11. P. Vianco, J. Rejent, I. Artaki, U. Ray, D. Finley, A. Jackson, "Compatibility of lead-free solders with lead containing surface finishes as a reliability issue in electronic assemblies," Electronic Components and Technology Conference, Orlando, FL, 1996, 1172–1183.
12. E. Bradley III, J. Hranisavlevic, "Characterization of the melting and wetting of Sn-Ag-X solders," 2000 Electronic Components and Technology Conference, Las Vegas, NV, 2000, 1443–1448.
13. Q. Zhu, M. Sheng, L. Luo, "The effect of Pb contamination on the microstructure and mechanical properties of SnAg/Cu and SnSb/Cu solder joints in SMT," Soldering and Surface Mount Technology, 12(3), 2000, 9–23.
14. J. Oliver, M. Nylén, O. Rod and C. Markou, "Fatigue properties of Sn3.5Ag0.7Cu solder joints and effects of Pb-contamination," Journal of SMT, 15(4), 2002, 23–29.
15. T. Ellis, "Tin whiskers," http://www.empf.org/empfasis/sept03/tinwisk1.htm.

Chapter 8
Component Reliability

Abstract In the industry, the most common way of predicting system reliability is to utilize empirical models. There are several standards-based approaches available. Many of them originate from the areas of industry, where high reliability has traditionally been of essence. Those include military and telecommunications. Empirical models are very popular, as they are easy to use and the standards do give quite comprehensive advice on how to use those. At system level, it is quite clear that the use of these simplistic models is more or less a standard procedure. The problem with empirical models is that they may not be equally well suited for component-level predictions. There, certain assumptions – such as constant failure rate and Arrhenius-type dependency (exponential) on temperature – that are used limit the applicability of these models. One should, however, remember that empirical models are usually applied to a large system, where many errors cancel out and the resulting reliability prediction is relatively accurate.

8.1 Introduction

Despite the fact that there has been evident progress in component quality and reliability [1], there are some signs of degradation of component reliability. One reason for this is the abandoning of the military handbooks that provided clear guidelines. Therefore, common requirements on acceptable reliability levels do not exist. Today's market is driven by consumer application-oriented systems instead of the ones that require long-lasting, high-reliability performance. This has sometimes resulted in a lack of components conforming to high-reliability requirements. This lack has caused some problems, especially in the application areas where long lifetime and high reliability are required, such as military [2] and telecommunications infrastructure products [3].

New surface mount component types without interconnection leads cannot always be adapted due to their limited reliability in demanding applications [4]. Several new component types have been introduced to the market, but the second-level interconnection reliability of all these components is not at a sufficient level. In Fig. 8.1, some thermal cycling test results are depicted [5]. It can be easily seen that most of the components do not conform to the no-failures-in-1,000-cycles criterion.

J. Liu et al., *Reliability of Microtechnology: Interconnects, Devices and Systems*,
DOI 10.1007/978-1-4419-5760-3_8, © Springer Science+Business Media, LLC 2011

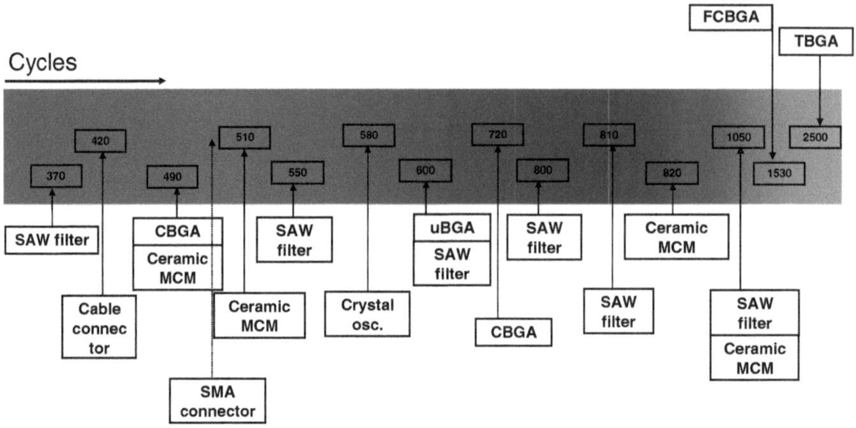

Fig. 8.1 Thermal cycling (−40 to +125°C, 1-h cycle) test results of some leadless components. Characteristic lifetimes in cycles are depicted [5]

The complexity of the products is increasing. This may also create further demands on component reliability. Outsourcing of the design and the manufacturing of IP blocks do not eliminate the responsibility of the end-product manufacturer. Outsourcing may even be seen as a threat to reliability and quality, unless the end-product manufacturer carefully communicates the reliability targets and controls the fulfillment of the reliability requirements.

In Chap. 3, general failure mechanisms are discussed, whereas in this chapter, the component reliability is looked at from an alternative point of view – empirical models.

8.2 Empirical Models

While physical models address a certain failure mechanism and try to give an estimate on lifetime based on the evolution rate of the degradation, empirical models are giving some generic estimates on failure rate for a certain component type or technology. Although being based on empirical data, the effect of field environment is taken into account by "factors" responsible for the degradation effects related to temperature, voltage, or some other stress factor. Therefore, these two ways of making lifetime estimates – physical models and empirical models – are not completely opposite, but both of them apply physical and chemical relations. In the case of empirical models, the actual failure mechanism is not, however, directly implicated, but more or less buried inside the model.

Empirical models have a long history and they are still widely applied. One of the major reasons for their popularity is the fact that they are relatively simple and easy to use. Also, when using empirical models, it is easy to expand the reliability analysis from component level to a system or a subassembly level. Many software tools also support the use of empirical models. Empirical models can, in principle, also take into account early failures and random failures, which is not usually the case when considering physical models.

Since the early 1970s, the failure rates for microdevices have fallen ca. 50% every 3 years [6], and the empirical models have been updated on the average every 6 years, thus, the models have become overly pessimistic. In 1994, the US Military Specifications and Standards Reform initiative led to the cancelation of many military specifications and standards, including MIL-HDBK-217 [7]. However, this was not the end of the story. PRISM and 217Plus are the updated versions, the old military handbook prepared by the Reliability Information Analysis Center (RiAC). Due to the popularity of MIL-HDBK-217, the Defense Standardization Program Office (DSPO) decided to revitalize MIL-HDBK-217. On May 8, 2008, the initial 217WG meeting was held in Indianapolis. At this time, the updating is on-going with the help of volunteering industry partners.

Besides MIL_HDBK-217, there are several other standards based on empirical models, such as Bellcore Reliability Prediction Procedure (Telcordia) [8], Nippon Telegraph and Telephone (NTT) procedure [9], British Telecom Handbook [10], CNET procedure [11], and Siemens procedure [12]. The predicted failure rates originating from different standards may, however, deviate from each other [6, 13].

8.3 The Methodology

Although each empirical model is a bit different from each other, there are several similarities between the models, and the basic methodology is quite similar. For each component technology, a certain base failure rate λ_b is defined. This failure rate is considered to be a typical or average failure rate representative for this specific component technology. The value for this failure rate is chosen based on the field failure data.

Base failure rate alone is rarely used, but it is usually multiplied by the so-called pi-factors that may take into account several factors: operational conditions (temperature π_T and voltage π_v), quality of the component π_Q, "learning factor" (based on the age of the component/technology) π_L, and "environmental factor" (taking into account the ambient conditions of the device use) π_e. The end result is the failure rate prediction for a certain component λ:

$$\lambda = \lambda_b \prod_{i=1}^{n} \pi_i, \tag{8.1}$$

One should note that even though the formulae may resemble each other, the parameter values, base failure rate λ_b and pi-factors π_i for different empirical models, may vary a lot, as well as the actual failure rate prediction λ.

Usually, the reliability prediction using empirical models is started in an early phase of product development. Then, only limited information on the actual design is available. Therefore, at this time quite often the effect of stress factors may be neglected. This kind of analysis is called *parts-count method*. When the electrical design gets more mature, more information becomes available and, therefore, the effect of voltage and temperature can also be better taken into account. At this stage, the methodology is called parts-stress.

8.4 Empirical Models in System Reliability Analysis

As was implied in Sect. 7.3, the empirical models give a constant failure rate for a certain component. This may not always be a realistic assumption, because this is, strictly speaking, valid only in the case of the so-called useful life period of a component's lifetime. This is the "middle part" of the bathtub curve, after the early-failure period and before the wear-out period.

However, it can be shown that reasonable approximations are available to turn nonconstant failure rates into quasi-constant values, as will be discussed in Chap. 9. Furthermore, using the constant failure rate assumptions makes the system-level reliability analysis very simple. This is due to the fact that the reliability function for a component that has a constant failure rate can be expressed as:

$$R(t) = e^{-\lambda t}. \tag{8.2}$$

When assuming that there are two components in a system having failure rates λ_1 and λ_2, the reliability function for this system (assuming that both components are required to be functional in order for the system to be operational $\longleftrightarrow$ series connection) can be given as:

$$R(t)_{sys} = R(t)_1 \cdot R(t)_2 = e^{-\lambda_1 t} \cdot e^{-\lambda_2 t}, \tag{8.3}$$

which is equivalent to:

$$\lambda_{sys} = \lambda_1 + \lambda_2. \tag{8.4}$$

For a system consisting of n components, the same can be written as:

$$\lambda_{sys} = \sum_{i=1}^{n} \lambda_i, \tag{8.5}$$

As can be seen, to calculate the system reliability, it is enough to sum the failure rates of each individual component. To be able to write (8.5), one needs to assume that the failure rates of the components are statistically independent. In a general system this assumption can be difficult to make, since each component's reliability could be a complex function of time, stress level, etc.

Mean time to failure (MTTF) for a system is simply:

$$MTTF_{sys} = \frac{1}{\lambda_{sys}}. \tag{8.6}$$

One can, however, argue that it is not realistic to consider that all components are required to operate to be able to consider the system to be operational. This is a valid argument, while not all components may be equally critical – and sometimes some redundancy is – on purpose – created so that the system can operate even in the case of failure of a certain component or a subsystem.

If more complex scenarios are to be studied, then more powerful tools are needed. Those include, e.g., the Reliability Block Diagram (RBD) technique or Markov Chain analysis technique [14]. When using these methods, the mathematical analysis, however, becomes more complex and in many cases – especially in the case of repairable systems – a simulation method, such as Monte Carlo, needs to be utilized.

Using empirical models in conjunction with series-system assumption (resulting in easy mathematics) is, however, not mandatory, even though that is more or less a standard practice. In theory, nothing prevents the use of empirical component models as part of RBD analysis. In practice, however, software tools are often so organized that empirical reliability prediction and RBD are separate modules. Assuming constant failure rate is not mandatory either. Some alternative approaches do exist [15].

8.5 Limitations of Empirical Models and Recommendations on Use

As discussed earlier in Sects. 8.2 and 8.4, there are several drawbacks and limitations to the use of empirical models. The validity and novelty of data on which the models are based is one of the most severe ones. Due to the rapid development of component technologies, many empirical models – unless frequently updated – can become obsolete.

It may be argued that the use and the use environment – on which the empirical model is based – may be very different from the one the component is about to be applied to. Therefore, selecting a telecom standard-based model is a good idea, if your design is about to be used in a telecom application. A military standard model may not be equally good choice in that case.

The effect of all stress factors is not comprehensively taken into account when developing the models. For example, the effect of vibration is not visible in the models, even though this kind of stressing can be embedded in the field data on which the model is based.

Interconnections are not usually taken into account in empirical models, even though their effect on reliability is increasing both in absolute (surface mount technology is dominant, solder interconnection are getting smaller) and relative manner (semiconductor devices have become more rugged due to significant improvements in the manufacturing processes). There is, however, no reason why interconnection could not be satisfactorily taken into account when using empirical models. One just needs to insert a representative model into those.

Furthermore, the physical models and their parameters embedded in the models have been heavily criticized. As an example, the fact that only Arrhenius-type (exponential) absolute temperature dependency is utilized, even though it is well known that this dependency can be more complicated [16]. Another criticism is related to the parameter value selection, which may not always have been best possible.

One further word of caution is that as empirical models are based on field failure data, they may not be very suitable for new, radically different component types. However, components that have only minor deviations from a reliability perspective are potentially easy to analyze using empirical models for existing component types.

Regarding simplifications related to system-level analysis, the main argument that has been assuming constant failure rate and a series-connection type relation between components (all components are needed to keep the system operating) are not necessarily realistic.

In refs. [6] and [13], the different models and the reliability estimates obtained when using those are studied. Both studies show a very large deviation between results obtained when using alternative models. In ref. [6], an analysis of a single component is performed, whereas in ref. [13], a system consisting of several components is also studied. Nevertheless, in both bases significant deviations are obtained. In Table 8.1, the failure rates for a memory component are listed. It can clearly be seen that not only the absolute values vary a lot, but also the temperature dependency is quite different. This is due to the different selection of activation energy values.

The situation is unfortunately not very much better when considering whole systems. When studying six different circuit board assemblies, the deviation could be even as high as 500% (over-pessimism) (Fig. 8.2). However, in certain cases, the failure rate proved to be much lower than anticipated.

When looking at these predictions and the evident deviations from the observed failure rate values, one should, however, remember that predicting reliability always means working with models and parameters with considerable uncertainty. Therefore, the fact that reliability is highly dependent is unfortunately true – but not depending on the model type. Finding the right activation energy value is a common task – be the model either empirical or physical.

To obtain the best possible accuracy, when using empirical models, it is recommended that a company updates the parameters based on their own field data. Doing so, the data best reflect the use and use environment the components are likely to encounter. It is also recommended that interconnections are taken into account in

Table 8.1 Failure rates for a memory component at different temperatures (© 1992 IEEE) [6]

Procedure	20°C	40°C	60°C	80°C
Hermetic packaging				
Mil-HDBK-217 (stress)	77	216	582	1,495
MIl-HDBK-217 (parts count)	219	219	219	219
Bellcore RPP	45	140	378	910
NTT procedure	84	139	259	541
CNET procedure (stress)	353	631	1,156	2,140
CNET (simplified)	790	1,950	–	–
British telecom procedure	8	8	10	19
Siemens procedure	48	96	208	504
Non-hermetic packaging				
Mil-HDBK-217 (stress)	86	362	1,532	5,648
MIl-HDBK-217 (parts count)	380	380	380	380
Bellcore RPP	54	168	454	1,092
NTT procedure	216	410	827	1,844
CNET procedure (stress)	446	835	1,687	4,088
CNET (simplified)	790	1,950	–	–
British telecom procedure	8	8	16	33
Siemens procedure	69	149	341	845

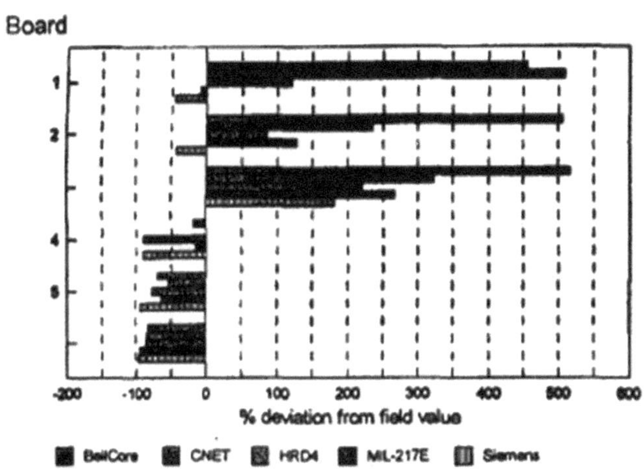

Fig. 8.2 Deviation from the observed failure rate for six different circuit board assemblies (© 1999 IEEE) [13]

the analysis phase, especially if the product is applying some novel interconnection technologies, where the risk of premature failure is larger.

As the primary use of empirical models is in the early phase of product development, a clever reliability engineer can greatly benefit if recognizing the potential risks early on. In an early phase, changes are still relatively easy to make. Therefore, the relatively poor accuracy may be compensated by the ease of use and possibility to be involved early in an R&D project. Quite often reliability models are used to compare different designs, and then the absolute accuracy of reliability predictions is not of primary importance, but the indication of primary risks is vital.

Exercises

8.1 Define the following terms (a) reliability, (b) availability, and (c) derating.
8.2 An equipment consists of a radio part (RF) and base band part (BB). Failure rate for RF part is 250 FIT and for the BB part 200 FIT (1 FIT = 1 failure/ 10^9 h). To operate, both parts need to be functional (RF and BB) (a) draw a Reliability Block Diagram (RBD), (b) calculate the failure rate for the whole equipment, (c) determine the reliability of the equipment after 10 years.
8.3 Draw a bathtub curve. Which three areas can be recognized and what they represent?
8.4 Compare the interconnection reliability of lead-free solders (like SnAgCu) to the interconnection reliability of SnPb. Describe how these materials behave in test and field environments. How the change of solder materials has affected different reliability prediction techniques?
8.5 List different ways to estimate reliability of a component. Describe the main quantitative methods to estimate reliability and give some examples on those.

References

1. D. L. Crook, "Evolution of VLSI Reliability Engineering", Proc. Int. Reliability Physics Symposium, 1990, 2–11.
2. E. Demko, "Commercial-Off-The-Shelf (COTS): A Challenge to Military Equipment Reliability", Proc. Annual Reliability and Maintainability Symposium, 1996, 7–12.
3. T. Ejim, "High Reliability Telecommunications Equipment: A All Order for Chip-Scale Packages", Chip Scale Review, 5, 1998, 44–48.
4. P. Frisk, M. Lindgren, B. Isaksson, P. Sneitz and K. Hagelin, "Novel Packages and Packaging Technologies for Use in Harsh Environments", Proc. of IMAPS Nordic, 2002.
5. O. Salmela, "Interconnection Reliability Scales", Nokia internal presentation, Nokia Networks, 2001.
6. J. B. Bowles, "A Survey of Reliability-Prediction Procedures for Microelectronic Devices", IEEE Transactions on Reliability, 41, 1992, 2–12.
7. MIL-HDBK-217, Reliability Prediction of Electronic Equipment, version E, 1986, October 27.
8. TS-TSY-000332, Reliability Prediction Procedure for Electronic Equipment, Issue 2, Bellcore, July 1988.
9. Standard Reliability Table for Semiconductor Devices, Nippon Telegraph and Telephone Corporation, March 1985.
10. Handbook of Reliability Data for Components Used in Telecommunications Systems, Issue 4, British Telecom, January 1987.
11. Recuiel De Donnes De Fiabilite Du CNET (Collection of Reliability Data from CNET), Centre National D'Etudes des Telecommunications (National Center for Telecommunication Studies), 1983.

12. SN29500, Reliability and Quality Specification Failure Rates of Components, Siemens Standard, 1986.
13. J. Jones, J. Hayes, "A Comparison of Electronic-Reliability Prediction Models", IEEE Transactions on Reliability, 48(2), 1999, 127–134.
14. D. T. P. O'Connor, Practical Reliability Engineering (Fourth Ed.), John Wiley & Sons, New York, 2002.
15. J. Moltoft, "Reliability Engineering Based on Field Information – The Way Ahead", Quality and Reliability Engineering International, 10, 1994, 399–409.
16. M. Pecht, "Why the Traditional Reliability Prediction Models Do Not Work – Is There an Alternative?", Electronics Cooling, 2(1), 1996, 10–12.

Chapter 9
System Level Reliability

Abstract System reliability is discussed in this chapter. In order to understand how the whole product or an entire system operates, it is necessary to combine the effects of individual components. One of the most commonly used approaches is the Reliability Block Diagram (RBD) methodology, where each component/element is represented by a block. The system consists of several blocks that are linked together based on their reliability criticality. For example, if all the blocks are in series, then all components need to operate for the whole system to operate. Parallel configuration may, however, operate even during a failure of one of its parallel components. First, the RBD methodology is discussed. Then, bathtub curve is introduced. After that, different options to approximate Weibull distribution in terms of constant failure rate are discussed, and some alternative approaches are benchmarked.

9.1 Introduction

System reliabilities can be calculated from individual component (or subsystem) reliabilities, if the series–parallel reliability relationships are known.

Series reliability refers to the situation where the system fails if any individual component, the weakest link, fails and is given for a system of n components by:

$$R_{SS} = \prod_{i=1}^{n} R_i = R_1 \cdot R_2 \cdot R_3 \cdot \cdots \cdot R_n, \tag{9.1}$$

with system hazard rate:

$$\lambda_{SS} = \sum_{1}^{n} \lambda_i, \tag{9.2}$$

and mean time to failure:

$$MTTF_{SS} = \frac{1}{\lambda_{SS}}. \tag{9.3}$$

J. Liu et al., *Reliability of Microtechnology: Interconnects, Devices and Systems*,
DOI 10.1007/978-1-4419-5760-3_9, © Springer Science+Business Media, LLC 2011

Table 9.1 System reliability reduction with complexity

	Individual component reliability			
Number of series components	99.999 (%)	99.990 (%)	99.900 (%)	99.000 (%)
10	99.99	99.90	99.004	90.44
100	99.90	99.01	90.48	36.60
250	99.75	97.53	77.87	8.1
500	99.50	95.12	60.64	0.66
1,000	99.01	90.48	36.77	0.004

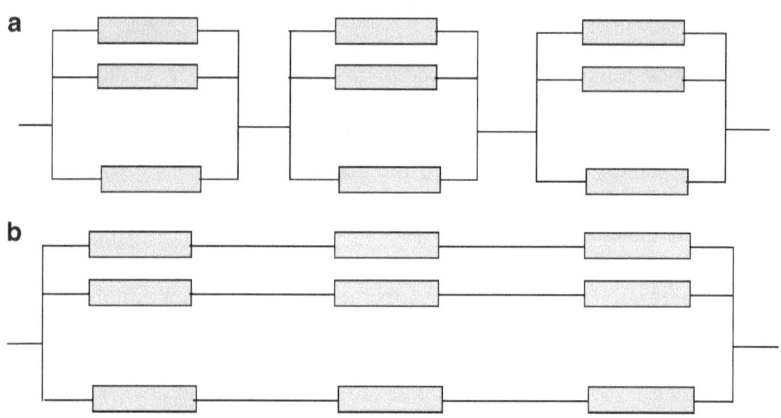

Fig. 9.1 (**a**) Series–parallel and (**b**) parallel–series reliability systems

Table 9.1 shows how rapidly system reliability can degrade in a series system of equal component reliabilities.

Parallel reliability applies to redundant systems and is given for a system of n components by:

$$R_{PS} = 1 - Q_{PS} = 1 - \prod_{i=1}^{n} (1 - R_i). \tag{9.4}$$

Formula for system hazard rate and MTTF escalate rapidly in complexity, so for $n = 2$, for example,

$$\lambda_{PS} = \frac{\lambda_1 e^{-\lambda_1 t} + \lambda_2 e^{-\lambda_2 t} - (\lambda_1 + \lambda_2) e^{-(\lambda_1+\lambda_2)t}}{e^{-\lambda_1 t} + e^{-\lambda_2 t} - e^{-(\lambda_1+\lambda_2)t}}. \tag{9.5}$$

Examples of the next level of complexity are shown in Fig. 9.1, which contrasts (a) the series combination of redundant elements with (b) a redundant arrangement of series elements. Real systems are often designed to include redundant combinations of low-reliability elements. More generalized series–parallel systems can be analyzed by a quasi-Boolean algebraic approach, as demonstrated here for the

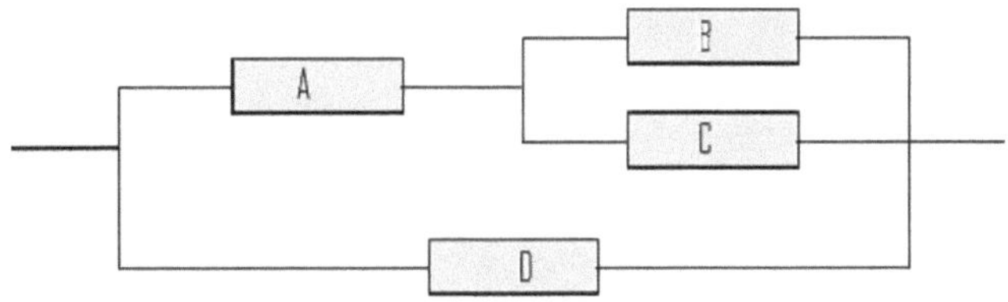

Fig. 9.2 Complex (mixed) reliability system

Table 9.2 Determination of system reliability

Operating states	Nonoperating states
$\mathrm{A.B.C.D}$	–
$\mathrm{A.B.\underline{C}.D}$	–
$\mathrm{A.\underline{B}.C.D}$	–
$\mathrm{A.\underline{B}.\underline{C}.D}$	–
$\mathrm{\underline{A}.B.C.D}$	$\mathrm{A.\underline{B}.\underline{C}.\underline{D}}$
$\mathrm{\underline{A}.B.\underline{C}.D}$	$\mathrm{\underline{A}.B.C.\underline{D}}$
$\mathrm{\underline{A}.\underline{B}.C.D}$	$\mathrm{\underline{A}.B.\underline{C}.\underline{D}}$
$\mathrm{\underline{A}.\underline{B}.\underline{C}.D}$	$\mathrm{\underline{A}.\underline{B}.C.\underline{D}}$
$\mathrm{A.B.C.\underline{D}}$	$\mathrm{\underline{A}.\underline{B}.\underline{C}.\underline{D}}$
$\mathrm{A.B.\underline{C}.\underline{D}}$	–
$\mathrm{A.\underline{B}.C.\underline{D}}$	–

system of Fig. 9.2. Table 9.2 lists all the possible operational (and nonoperational) conditions in terms of functional and nonfunctional elements, indicated for example by A and $\underline{\mathrm{A}}$ respectively. Considering all the possible operational states, the complex system reliability can be written in terms of the component reliabilities R_A, R_B, R_C, and R_D, as:

$$R = R_A R_B R_C R_D + R_A R_B (1 - R_C) R_D + R_A (1 - R_B) R_C R_D + \ldots \text{etc.}, \tag{9.6}$$

which can be simplified to show that:

$$R = R_D + R_A (R_B + R_C - R_B R_C)(1 - R_D). \tag{9.7}$$

It is well known that only the exponential distribution has a constant hazard rate. The constant hazard rate is related to some random effects that take place during the lifetime of a component (bathtub curve with $\beta = 1$ in Fig. 9.3).

When Weibull shape parameter $\beta < 1$, failures are predominantly of early failure type, and when $\beta = 1$, random failures are dominant, and when $\beta > 1$, wearout is mostly responsible for failures.

An exponential distribution assumption with constant hazard rate is used quite a lot due to the resulting simplicity in system level reliability analyses. When utilizing a constant hazard rate assumption in parts-count type reliability estimates,

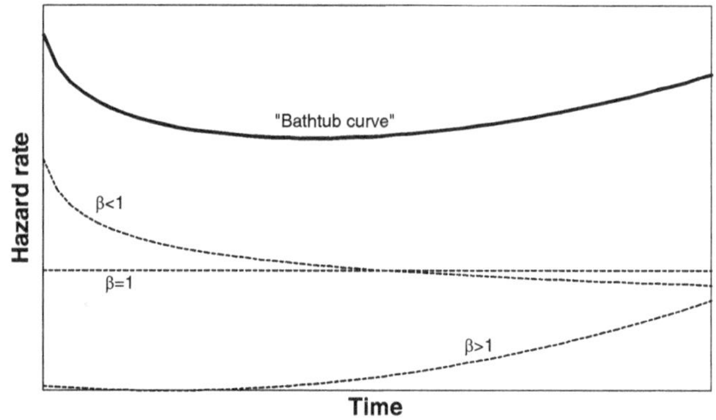

Fig. 9.3 Bathtub curve and the different failure regions

the hazard rates of individual components $\lambda_{comp,i}$ can be summed up, and the end result is the system level hazard rate λ_{system} [1]:

$$\lambda_{system} = \sum_{i=1}^{n} \lambda_{comp,i}. \tag{9.8}$$

The reciprocal of the system hazard rate is the MTTF (Mean Time to Failure) of the system:

$$MTTF = \frac{1}{\lambda_{system}}. \tag{9.9}$$

Quite a lot of component lifetime data that have been gathered are presented in terms of constant hazard rate. Many system level reliability prediction methods also give lifetime predictions in terms of constant hazard rates [2]. However, in reality, the constant hazard rate assumption is often not valid. Therefore, applying exponential distribution may not always be an appropriate choice [3]. Assuming a constant hazard rate makes the mathematical analyses easy, but assuming a constant hazard rate is in contradiction with the fact that most components fail either in the early failure or the wearout regime, where the hazard rate is either decreasing or increasing, respectively. The hazard rate in those regimes can be taken into account, for example, by utilizing Weibull statistics, but not by an exponential distribution. Owing to this fact, there seems to be an unbridgeable situation, as component level reliability data can be interpreted by applying Weibull statistics, but these results cannot be utilized later on in simplistic system level MTTF calculations.

The relationship between the exponential and the Weibull distributions has already been studied in the past, and the so-called Weibull-to-exponential transformation has been created [4–6]. The use of this transformation simplifies the estimation of the confidence bounds and some other parameters of the Weibull distribution. When using the transformation, the Weibull data is first transformed into an exponential form where the mathematical analyses, (e.g., the determination of the confidence bounds) are done. After that, the results are converted back to Weibull form.

In our case, the Weibull data (hazard rate) is converted into exponential type data format (constant hazard rate) by time-averaging the hazard rate within certain time intervals. The approximate information created is readily applicable in parts-count type system level reliability analyses. Conversion back to the Weibull regime is not needed.

9.2 Some Constant Hazard Rate Approximations of the Weibull Distribution

The exponential distribution and Weibull distributions are of different forms, and they have a different time dependency. The only exception is the case when the Weibull distribution shape parameter $\beta = 1$, in which case the two distributions are identical, with $\eta = \theta = 1/\lambda$. In this case, the Weibull distribution characteristic lifetime η is equal to the MTTF (θ) value of the exponential distribution. At all other times, the distributions are not identical and therefore, some approximation is needed in order to present the Weibull distribution data in terms of an exponential distribution.

There may be different strategies to create a suitable approximation of the Weibull distribution. Although it is impossible to match all the distribution functions (hazard function $h(t)$, probability density function $f(t)$, cumulative density function $F(t)$, and reliability function $R(t)$) between the two distributions simultaneously, there is a possibility to match perfectly some individual functions.

After the two-parameter Weibull data are transformed into constant hazard rate form, it can be utilized in MTTF calculations for the whole system. Therefore, it would be beneficial if the reliability function of the approximate exponential distribution $R(t)_{WB\rightarrow EXP}$ would imitate the reliability function of the original Weibull distribution $R(t)_{WB}$ as closely as possible, in other words:

$$R(t)_{WB\rightarrow EXP} \approx R(t)_{WB}. \tag{9.10}$$

Another criterion to be fulfilled is that the form of the hazard function $h(t)_{WB\rightarrow EXP}$ should be kept as simple as possible, but it should still present the main characteristics of the original distribution. This means that preferably $h(t)_{WB\rightarrow EXP} =$ *constant at least for some time intervals*. Still, oversimplification should be avoided when trying to satisfy this criterion. Otherwise, some false conclusions might be drawn from the MTTF calculations. Typically, the reliability test results of components are of increasing hazard rate type.

Weibull distribution with two parameters, shape parameter β and the characteristic lifetime η, can fit the data satisfactorily many times.

The Weibull hazard rate is of the form [7]:

$$h(t) = \beta \cdot t^{\beta-1} / \eta^{\beta}. \tag{9.11}$$

In order to approximate this function, one of the below strategies can be chosen:

Option 1: Pick some representative value of the hazard function at some selected time t.
Option 2: Calculate a time-averaged hazard rate value for the whole lifetime.
Option 3: Calculate a time-averaged hazard rate value for some time intervals.
Option 4: Pick values from the time-averaged hazard rate curve (Option 2) between selected time intervals.
Option 5: Calculate time-averaged reliability function values for selected time intervals and based on those, calculate equivalent hazard rate values λ_{eq} for each time interval.

The actual procedure is explained later on in more detail.

In the following section, the five strategies above are discussed in light of the criteria given earlier in this chapter.

First, give the formal definitions for Options 2–5:

- Option 2
  The hazard rate of the option 2 is defined as the time-averaged value over the whole lifetime of the component:

$$\langle h(t) \rangle_t = \frac{\int_0^t h(t')\mathrm{d}t'}{\int_0^t \mathrm{d}t'} = \frac{t^{\beta-1}}{\eta^{\beta}}. \tag{9.12}$$

  It is noted that this value is dependent on time t. The above approximation is useful, if the expected lifetime or lifetime requirement for the component $t = t_{lifetime}$ is known. By inserting this value into (9.12), it results in one constant hazard rate value for the whole lifetime of the component.
- Option 3
  The third option can be calculated in a similar way as above, but this time, the time-averaged hazard rate will be calculated for selected time intervals $\Delta t = t_{i+1} - t_i$:

$$\langle h(t) \rangle_{\Delta t} = \frac{\int_{t_i}^{t_{i+1}} h(t)\mathrm{d}t}{\int_{t_i}^{t_{i+1}} \mathrm{d}t} = \frac{1}{\eta^{\beta}} \cdot \frac{\left(t_{i+1}^{\beta} - t_i^{\beta}\right)}{t_{i+1} - t_i}. \tag{9.13}$$

  In this case, the hazard rate has a constant value in a selected time interval from t_i to t_{i+1} $i = 0, 1, 2, \ldots, n$, where n is the number of time intervals.
- Option 4
  This option makes use of time-averaged hazard rate function defined by (9.12). The hazard rate values used are defined as $\langle \mathrm{h}(t_{i+1}) \rangle_t$ during selected time intervals:

$$\Delta t = t_{i+1} - t_i$$

- Option 5
  Utilizing Option 5 requires a little more rigorous analysis. The strategy is to first solve the time-averaged value of the reliability function RWB for selected time intervals $t_i \ldots t_{i+1}$.
  This can be accomplished by writing:

$$\langle R_{\mathrm{WB}} \rangle = \frac{\int_{t_i}^{t_{i+1}} R(t)dt}{\int_{t_i}^{t_{i+1}} dt} = \frac{\int_{t_i}^{t_{i+1}} \mathrm{e}^{-(t/\eta)^\beta} \mathrm{d}t}{t_{i+1} - t_i}$$
$$= \frac{\eta}{\beta(t_{i+1} - t_i)} \left\{ \Gamma\left(\frac{1}{\beta}, \left(\frac{t_{i+1}}{\eta}\right)^\beta\right) - \Gamma\left(\frac{1}{\beta}, \left(\frac{t_i}{\eta}\right)^\beta\right) \right\}, \tag{9.14}$$

where $\Gamma(\cdot,\cdot)$ is the incomplete gamma function. In Fig. 9.4, the time-averaged reliability function is depicted.

The instant in time t_{eq} ($t_i \leq t_{eq} \leq t_{i+1}$), at which the time-averaged reliability function is equal to the reliability function of the original Weibull distribution, may be written as:

$$t_{eq} = \eta \left[\ln\left(\frac{1}{\langle R_{WB} \rangle} \right) \right]^{1/\beta}. \tag{9.15}$$

In order to obtain the corresponding equivalent constant hazard rate λ_{eq}, the exponential reliability function R_{EXP} can be utilized:

$$R_{EXP} = e^{-\lambda_{eq} t}. \tag{9.16}$$

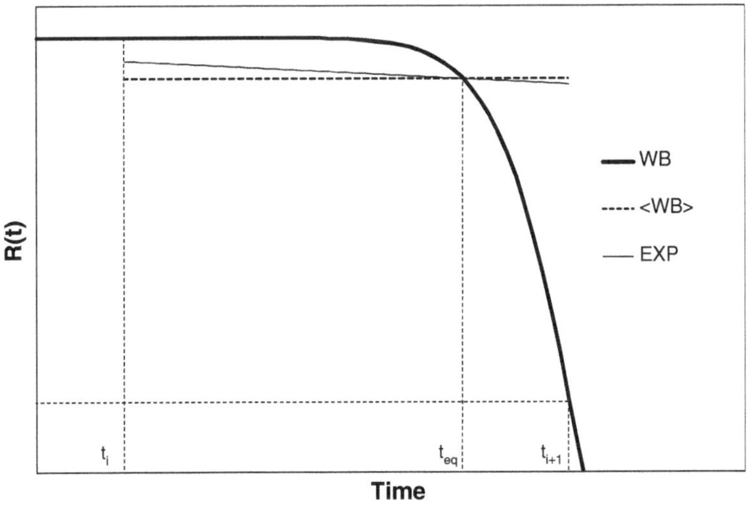

Fig. 9.4 The Weibull reliability function $R(t)$ (WB), the time-averaged reliability function R_{WB} ($\langle$WB$\rangle$), and the approximate exponential reliability function R_{EXP}(EXP) for time interval $t_i \ldots t_{i+1}$

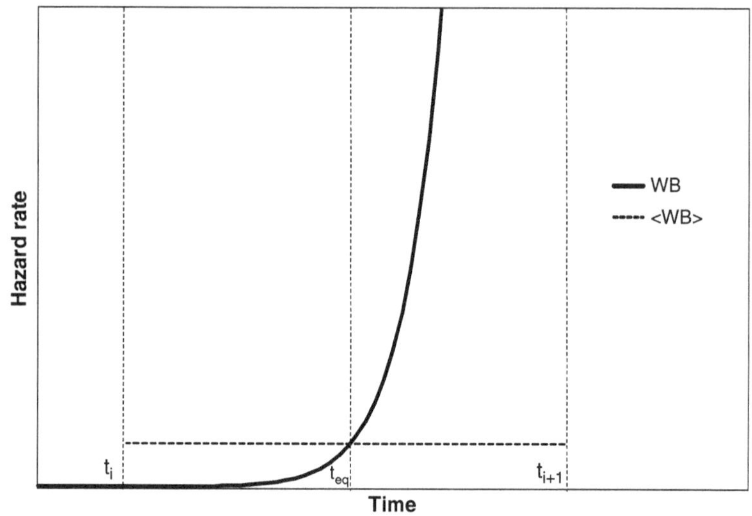

Fig. 9.5 Hazard rate of Weibull distribution (WB) and the time-averaged value ($\langle$WB$\rangle$)

To satisfy (9.10), it can be required that when $t = t_{\text{eq}}$, $R_{\text{EXP}} = R_{\text{WB}}$. After solving for λ_{eq}, the following is obtained:

$$\lambda_{eq} = \frac{\ln(1/\langle R_{WB} \rangle)}{t_{eq}}. \tag{9.17}$$

In Fig. 9.5, the Weibull and time-averaged hazard rate λ_{eq} are depicted.

Later on, it is shown that Option 5 best fulfills the requirement given by (9.10).

However, it may be demanding to calculate numerically the incomplete gamma function values accurately when time has large values, especially if β is large. In general, this is due to the lack of numerical solutions that are accurate enough for the incomplete gamma function, when variables have very large values.

9.3 Resulting Functions and Hazard Rates

In Fig. 9.6, all five approximate hazard rate options depicted for a component having $\eta = 3{,}677$ days and $\beta = 20$ can be seen. The time interval selected in the time averaging was 5 years. The hazard rate for Options 3–5 is, therefore, constant in time intervals 0...5 years, 5...10 years, 10...15 years, and 15...20 years.

The hazard rate for Option 1 is selected to be 10,000 FITs corresponding to the hazard rate value of Weibull distribution in the middle of the lifetime (10 years = 20 years/2). However, some other choice might have been justified as well. The hazard rate for Option 2 is the time-averaged value for the whole 20-year lifetime obtained by utilizing (9.12). The hazard rate for Option 3 was obtained by utilizing (9.13) with time-interval $t_{i+1} - t_i = 5$ years. Values for Option 4 are picked from

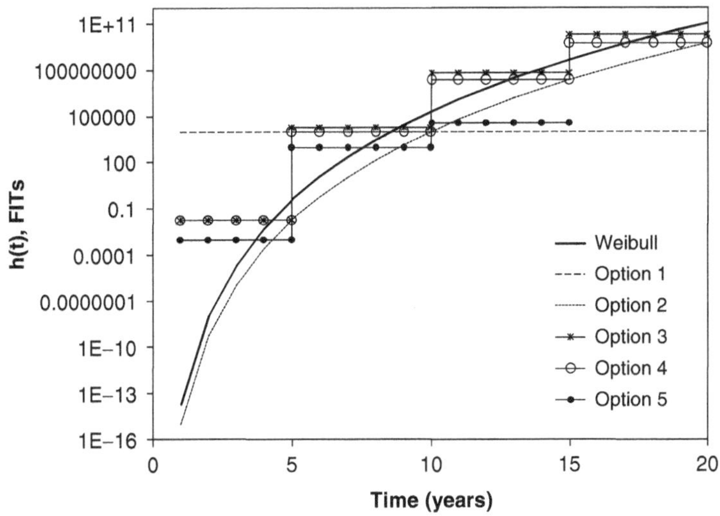

Fig. 9.6 Weibull hazard rate and five approximate options. The selected time interval used in time averaging is 5 years

the curve plotted according to (9.12) at time instants of 5, 10, 15, and 20 years. The hazard rate for Option 5 is calculated by utilizing the above-described method (9.13)–(9.17), which is based on the time averaging of the reliability function.

It is noted that the actual hazard rate obtains values from 0 FIT to 1×10^{11} FIT during the component's lifetime. Therefore, it might not be a good idea to use one single hazard rate value, as is the case in Option 1. If doing so, there is a danger that the value picked is not representative of the risk level of the component at all instants of time. Also, utilizing Option 2 with only one single hazard rate value results in a similar problem, although in this case the selection of the hazard rate is not arbitrary.

Keeping in mind the criterion stated in (9.10), the reliability function of the different options (Fig. 9.7) should also be studied. Doing so, it can be noted that a perfect fit between the original Weibull reliability function and Option 2 exists. The next best choices are Options 5, 4, and 3. Option 1 has the worst performance. Therefore, it is not a suitable choice.

If the exact lifetime expectancy $t_{lifetime}$ of a component were known prior to the product launch, then Option 2 would match exactly the original Weibull reliability function at $t = t_{lifetime}$. In this case, one would just pick $h(t_{lifetime})$ and use that in the MTTF calculations. This would represent the time-averaged value over the whole lifetime. However, in practice the true expected lifetime is not always known. Moreover, if wearout is expected to take place during the operational lifetime, averaging over the whole lifetime may result in a very large hazard rate value. This would not give a proper picture of the reliability of the component during its early life period. Therefore, Option 2 is attractive only if the hazard rate does not change much during the lifetime of a component. Keeping in mind that:

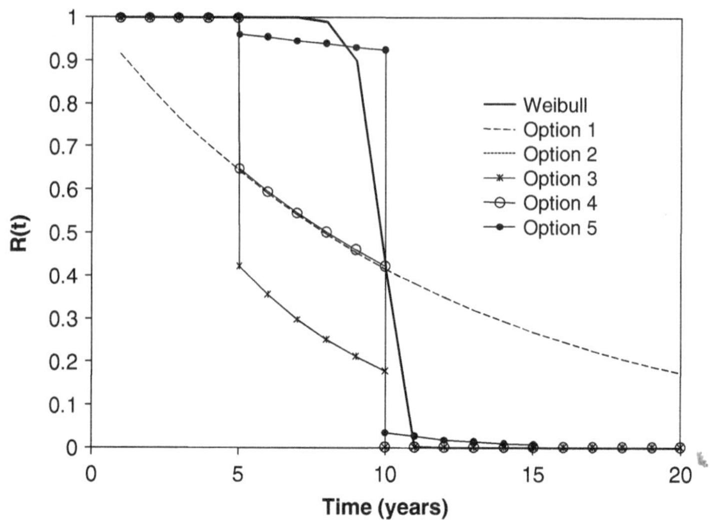

Fig. 9.7 Reliability functions of the different approximation options. Option 2 data is overlapping with the Weibull data. The time interval used in the time averaging is 5 years

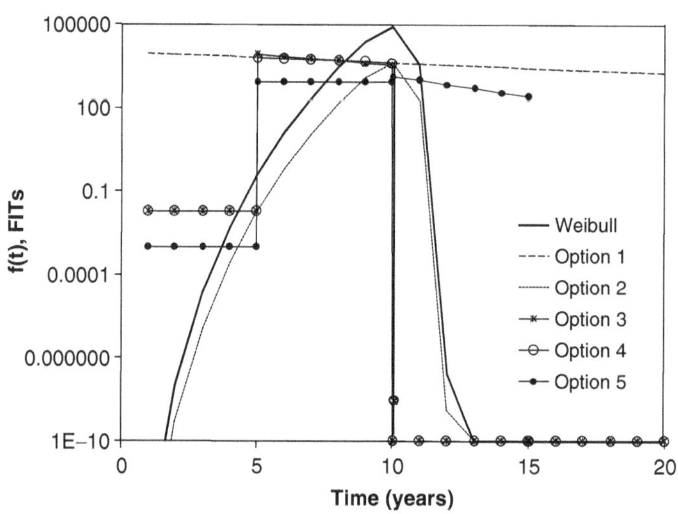

Fig. 9.8 Reliability density function of the Weibull and those related to the approximate solutions

$$F(t) = 1 - R(t). \tag{9.18}$$

It is expected that the approximate options behave similarly when cumulative failure function $F(t)$ is concerned.

Table 9.3 Time-averaged hazard rate values for different approximate options

Time	Approximate hazard rate (FITs)					
(years)	Weibull	Option 1	Option 2	Option 3	Option 4	Option 5
0...5	0...0.4	10,000	0...0.2	0.02	0.02	0.001
5...10	0.4...200,000	10,000	0.2...10,000	20,000	10,000	899
10...15	200,000...4 × 10^8	10,000	10,000...22 × 10^6	65 × 10^6	22 × 10^6	37,857
15...20	4 × 10^8...10^{11}	10,000	22 × 10^6...5 × 10^9	20 × 10^9	5 × 10^9	N/A

Looking at the density function $f(t)$, it may be noted that all the approximate solutions are a poor fit for the original Weibull distribution function (Fig. 9.8). One can also show, that:

$$\int_0^\infty f(t)dt < 1, \tag{9.19}$$

in the case of Options 2–4. Therefore, those options cannot be considered as true statistical distribution functions. The integration of a true distribution density function over time should always be equal to 1 [8].

When using Options 3–5, simple constant hazard rate values can be found for some selected time intervals, for example, in a tabulated form. This is demonstrated in Table 9.3 where the data of the above example is listed. Using Option 4 does not gain a hazard rate value during time interval 15...20 years due to the lack of accurate numerical solutions to the incomplete gamma function, as discussed earlier.

This kind of data can be utilized directly in parts-count type system level MTTF calculations.

9.4 Properties of Different Options

Let us first look at Option 2 in detail. The definitions of the statistical functions of Option 2 are based on the exponential distribution function using the hazard rate obtained from (9.12). This is accomplished just by replacing the constant hazard rate value λ by the hazard rate value given by the above definition (9.12). The functions of the exponential distribution and Option 2 are listed below in Table 9.4. The distribution functions derived for other options were also derived by replacing the exponential hazard rate function with the time-averaged hazard rate values.

As already shown, the reliability function of Option 2 is equal to the original Weibull reliability function at any selected instant in time t. Simple relations can be written between all statistical functions of the two-parameter Weibull distribution and those of Option 2. Table 9.5 lists these relations. Inserting the hazard rate defined by (9.19) into Option 2 distribution functions in Table 9.2 can verify that the relations are correct.

Table 9.4 Exponential distribution functions and Option 2 related functions

	Statistical distribution functional/statistical function value	
	Exponential	Option 2
Hazard rate	$h(t) = \lambda$	$h(t) = \langle h(t) \rangle_t$
Distribution function	$f(t) = \lambda e^{-\lambda t}$	$f(t) = \langle h(t) \rangle_t e^{-\langle h(t) \rangle_t t}$
Cumulative distribution function	$F(t) = 1 - e^{-\lambda t}$	$F(t) = 1 - e^{-\langle h(t) \rangle_t t}$
Reliability function	$R(t) = e^{-\lambda t}$	$R(t) = e^{-\langle h(t) \rangle_t t}$

Table 9.5 Statistical functions of the Weibull distribution, and their relationship to those of Option 2

	Statistical distribution functional/statistical function value	
	Weibull	Option 2, in terms of Weibull distr.
Hazard rate	$h(t) = \frac{\beta t^{\beta-1}}{\eta^\beta}$	$\beta h(t)$
Distribution function	$f(t) = \frac{\beta}{\eta^\beta} t^{\beta-1} e^{-(t/\eta)^\beta}$	$\beta f(t)$
Cumulative distribution function	$F(t) = 1 - e^{-(t/\eta)^\beta}$	$F(t)$
Reliability function	$R(t) = e^{-(t/\eta)^\beta}$	$R(t)$

An important note is that although closed form results can be derived for Option 2, Option 2 is not a true distribution function, as it does not satisfy all the criteria required from a true reliability statistical function (9.19). Actually, it can be shown that the integration of this function, over time, is equal to $1/\beta$. This may sound a bit odd, as both the cumulative distribution function and the reliability function for Option 2 get reasonable values and reach values in the whole scale (0…1). The explanation for this apparent contradiction is simply the fact that the cumulative distribution function, in this case, is defined by making use of the exponential function – not by actually integrating the distribution density function of the Option 2 over time.

Option 3 fitted both to hazard rate and reliability functions of the true Weibull distribution (Figs. 9.6 and 9.7) relatively accurately. Looking more carefully at the hazard rate function of this option, it is noted that at the end of the first time interval, the value of the hazard rate function is equal to the time-averaged value of the hazard rate (Option 2). During the next time intervals, the hazard rate of Option 3 starts to approach the original (instantaneous) Weibull distribution hazard rate. In actual fact, it can be shown that when the number of time intervals n approaches infinity, the hazard rate functions of Option 3 and the instantaneous Weibull distribution approach each other. The reliability function of Option 3 has always got smaller values than the true Weibull distribution (Fig. 9.7).

Option 4 is making use of the time-averaged hazard rate function defined by (9.12) at the end points of the time intervals. The reliability function is smaller than, or equal to, the original Weibull distribution function at all instants in time. At the end points of the time intervals, the reliability function is equal to the values given by the Weibull distribution and is smaller elsewhere. Option 4 is a better match to the original Weibull reliability function than Option 3.

Option 5 resembles most the original Weibull reliability function among those approximations that utilize time intervals. However, for very large time values, the calculation of the hazard rate may become cumbersome due to numerical solution accuracy limitations discussed earlier.

9.5 Comparison of the Selected Options

There are at least two things that must be taken into account, when making practical choices about the hazard rate approximation function. The first one is that the reliability function of the approximation should closely imitate the original Weibull reliability function. Option 2 is superior to the others in this respect as it matches perfectly the original Weibull reliability function. The next best choices are Options 5, 4, and 3. The use of a single, constant hazard rate value (Option 1) has the worst accuracy over the lifetime.

The other important criterion is to keep the expression of the hazard rate as simple as possible. By doing so, it is possible to apply the calculated hazard rate values directly into the system level parts-count type MTTF calculations. In this respect, Option 2 might not be a suitable choice, as it cannot be used in a tabulated form. All other options can be presented in a simple table form having constant hazard rate values either for the whole lifetime or for part of it.

To satisfy both criteria, Option 5 seems to be the best choice, having the possibility to be used in a simplistic form (for example, table) and still match reasonably well the true reliability behavior of the component.

9.6 Selection of Time Intervals

When using the simplistic time-averaged hazard rates, the time intervals should be selected in a way that the reliability behavior can be imitated with acceptable accuracy. In order to be able to satisfy this criterion, the reliability function should be plotted in conjunction with the hazard rate of the component and then the lifetime should be divided into suitable time intervals. There should be at least one, but preferably several, time intervals in which wearout has not yet fully occurred (let us say, $F(t) < 1\%$). The following time intervals may already include the wearout phenomena related to high hazard rate values, and therefore, the resulting time-averaged hazard rate value may be large in those intervals. When wearout has occurred almost completely, the hazard rate gets values of infinite magnitude and using those in the MTTF calculations will result in a clear message; this component will fail at latest in the selected time interval. One interval indicating the end of the life of the component is enough for practical purposes.

9.7 The Motivation for Selecting Two-Parameter Weibull Distribution

In this chapter, the two-parameter Weibull distribution was selected to present the statistical behavior of components that face wearout phenomena. Some other choice might have been possible, too. The selection of a suitable statistical distribution has raised some discussion in the science community. In [9] the two-parameter, Weibull distribution is recommended, whereas in [10, 11] the three-parameter, Weibull is considered superior over two-parameter Weibull. Also, lognormal distribution is considered to fit the test results better than two-parameter Weibull distribution. The conclusion that two-parameter Weibull distribution is not very accurately presenting the test data is based on least squares curve fitting results and the related small correlation coefficients obtained when fitting the test data to two-parameter Weibull distribution.

Another argumentation used against the two-parameter Weibull distribution is that it is expected that there is a failure-free period of time (presented by the failure-free time γ in the three-parameter Weibull distribution) when testing solder attachments. One fact supporting this is that according to Darveaux [12], it takes some finite time to initiate a crack in the solder material. One further observation made is that when fitting the test data to a two-parameter Weibull distribution, the test data has a tendency to have a downward sloping in the beginning of the wear-out period [10]. This is believed to indicate that there is a failure-free time that a two-parameter Weibull distribution cannot satisfactorily take into account. Furthermore, it is noted that if a two-parameter Weibull distribution is used, the reliability requirement based on it will be very demanding [10, 11].

Now, we try if we can verify that the two-parameter Weibull distribution is accurate enough for practical purposes. The author is aware that using the two-parameter Weibull distribution will result in a more demanding reliability requirement if very small percentages of failed items are considered. This is evident if comparing the behavior of cumulative distribution functions. It is also "natural" to consider that there is a failure-free period of time until the first items start failing in the test. However, we think that in reality, it is not impossible that items may fail very early. This may happen if the test vehicles are inherently very weak or if the test itself is very harsh. One should remember that as lifetime is often monitored in terms of number of cycles, this measure used is discretized, as the length of thermal cycle is finite. The first cycle may include the incubation period of some weak components. Still, from a number-of-cycles viewpoint, it would seem that the failure occurs instantly.

Therefore, the assumption of an incubation period is not necessarily in conflict with the selection of the two-parameter Weibull distribution. Furthermore, it is not known that there would be well-documented tests that would prove either two-parameter or three-parameter Weibull statistics to best describe the behavior of a test population, especially when very small cumulative failure percentages, such as 0.01%, are considered. This would require testing of thousands of items, which is very difficult to arrange in practice. Therefore, the discussion on the distribution function selection is at least partly speculative, as no actual proof exists.

9.8 Constant Failure Rate and Its Origin in the Field Failure Data

In the field environment, constant hazard rate at the product level is often recorded, although components may fail due to wearout phenomena. The reason that the exponential portion of the bathtub curve for a population of products is observed is in part because of repairs, and in part because of random overstress events through the lifetime of the population. If the data is grouped by failure mechanisms, then it is highly doubtful to find an exponential distribution for each group. It is more likely to find a collection of Weibull distributions, each with $\beta \neq 1$, indicating that either early failures or wearout mechanisms are taking place. However, at the system level, this can be represented with an averaged quasi-constant hazard rate.

Exercises

9.1–9.4 Calculate reliability for the following topologies:

9.1

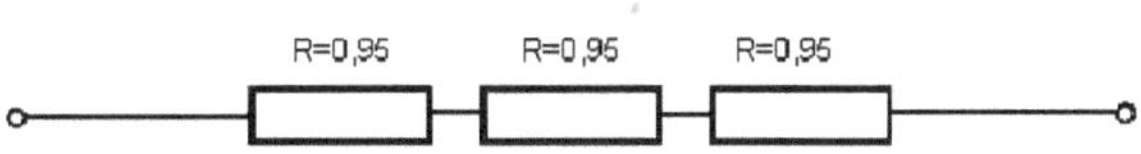

9.2

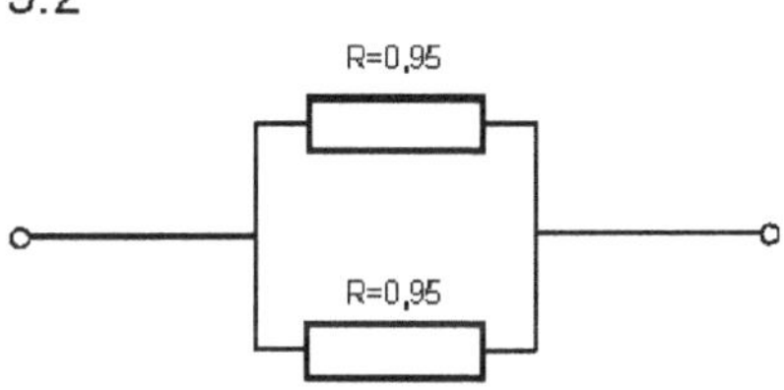

9.3

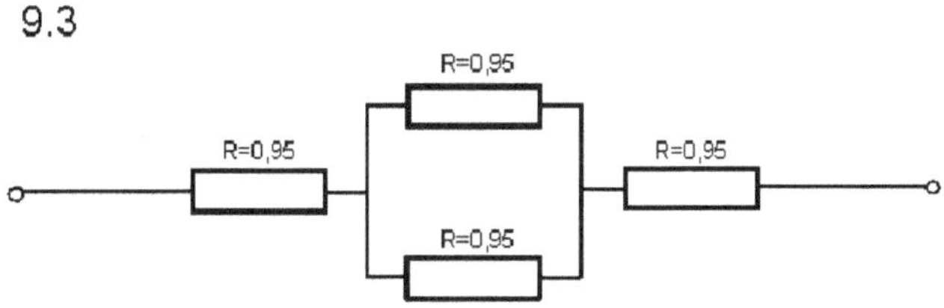

9.4

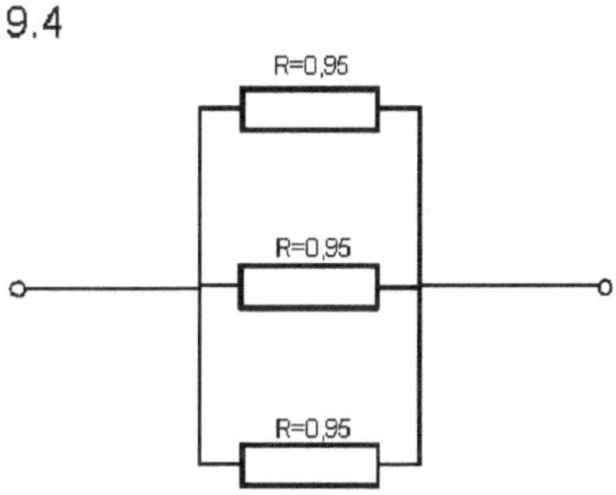

9.5 When should we use Weibull distribution, and when should we use log-normal distribution?

References

1. MIL-HDBK-217, Reliability Prediction of Electronic Equipment, Version E, October, 27, 1986.
2. J. B. Bowles, "A Survey of Reliability-Prediction Procedures For Microelectronic Devices," IEEE Transactions on Reliability, 41, 1992, 2–12.
3. J. B. Bowles, "Commentary-Caution: Constant Failure-Rate Models May Be Hazardous to Your Design," IEEE Transactions on Reliability, 51, 2002, 375–377.
4. W. Nelson, "Weibull Analysis of Reliability Data with Few or No Failures," Journal of Quality Technology, 17, 1985, 140–146.
5. J. B. Keats, P. C. Nahar, and K. M. Korbel, "A Study of the Effects of Mis-Specification of the Weibull Shape Parameter on Confidence Bounds Based on the Weibull-to-Exponential Transformation," Quality and Reliability Engineering International, 16, 2000, 27–31.
6. M. Xie, Z. Yang, and O. Gaugoin, "More on the Mis-Specification of the Shape Parameter with Weibull-to-Exponential Transformation," Quality and Reliability Engineering International, 16, 2000, 281–290.
7. P. D. T. O'Connor, Practical Reliability Engineering, John Wiley & Sons, New York, 1991.
8. W. Nelson, Accelerated Testing, Statistical Models, Test Plans, and Data Analyses, John Wiley & Sons, New York, 1990.
9. Guidelines for Accelerated Reliability Testing of Surface Mount Solder Attachments, IPC-SM-785, November 1992.
10. E. Nicewarner, "Historical Failure Distribution and Significant Factors Affecting Surface Mount Solder Joint Fatigue Life," Proc. International Electronic Packaging Conference, Wheaton, USA, 1993.
11. J. M. Clech, D. M. Noctor, J. C. Manock, G. W. Lynott, and F. E. Bader, "Surface Mount Assembly Failure Statistics and Failure Free Time," Proc. 44th Electronic Components and Technology Conference, May 1994.
12. R. Darveaux, "Effect of Simulation Methodology on Solder Joint Crack Growth Correlation," Proc. of the 50th Electronics Components & Technology Conference, Las Vegas, 2000, 1048–1058.

Chapter 10
Reliability and Quality Management of Microsystem

Abstract The ability to design and manufacture microsystems cost-effectively with high reliability and with short time-to-market is crucial for a company's competitiveness. To avoid delays in product release, it is necessary to focus on minimizing the risk for reliability problems by identifying reliability issues early in the design phase and designing-in features that assure reliability. This requires that failure modes that may be crucial for the reliability must be identified and that measures must be taken to mitigate associated failure mechanisms. Input required for identification of crucial failure mechanisms is data about product requirements, life-cycle conditions, architectures, and manufacturing processes. All involved in the product development process including the end customer must be involved in this work.

When new technologies are implemented, it is the product architecture and processes rather than the end product that shall be qualified. Factors that may affect reliability during production must be identified, and process control must be implemented to assure low variance in the production process.

10.1 Introduction

The Department of Defense in the USA and other standardization bodies, as well as many organizations and companies, have pointed out a performance-based approach as the only realistic alternative to the traditional standards-based approach. However, changing to a performance-based approach is not a question of replacing one set of standards with a new set of standards and implementing a few new tools. Those who wait for a new set of "how-to" standards will be disappointed because the performance-based approach is based on the realization that it is not possible to assure quality through that type of standards. There will be a few general standards, but most documents will be in the form of guidelines giving advice on activities that need to be carried out to assure quality.

The objective of performance-based quality management is to assure that the customer's expectations are met. It will be up to a manufacturer to make the final decisions, in cooperation with the customer, of how quality shall be assured. The customer's expectations must be captured and transformed into a form that

J. Liu et al., *Reliability of Microtechnology: Interconnects, Devices and Systems*,
DOI 10.1007/978-1-4419-5760-3_10,

can be used to assure the quality during the design and manufacture of a product. This is achieved by formulating the requirements of the product into a product specification. Functional requirements have always been handled like that. In that respect, it is not a new approach. What is new in the performance-based approach is that it is acknowledged that assurance of the quality of electronic hardware must be handled analogously to functional requirements. That is, all requirements need to be application-specific and performance-based.

Since functional requirements are already treated with a performance-based approach in current practice, this chapter focuses on the consequences of a performance-based approach for assurance of other quality issues. The main emphasis is on how to assure reliability and manufacturability; also, assurance of testability, maintainability, environmental compatibility, etc. are covered to some extent.

Assessment of reliability is more correctly described as assessment of unreliability, since it is the failure rate or probability of failure that is determined. In the standards-based approach, the failure rate is determined from Mean Time Between Failures (MTBF) values for the components making up the system. The MTBF values of the components are calculated from field data. There is no need for understanding of the failure mechanisms when predicting the reliability using this approach. Furthermore, the reliability is not considered to be affected by design and manufacturing processes and only to a small extent by field conditions, and then mainly the steady-state temperature of some components.

The performance-based approach is based on the philosophy that adequate assessment of failure rates should, as far as possible, be based on knowledge of the root causes of the failure mechanisms. For this reason, this approach is also called the physics-of-failure approach. It is sometimes claimed that this approach is only applicable for wearout failures, but there is no reason for such a limitation. Knowledge of the root causes for failures due to defects or overstress is important both for assessment and improvement of the reliability.

The failure mechanisms must be known for all crucial failure modes for a complete assessment of the reliability of a product. It is recognized that the failure mechanisms are affected not only by design (choice of materials and product architecture) and manufacturing processes but also by the conditions that the product will be exposed to during its entire life. Thus, the impact of design, manufacturing processes, and life-cycle conditions must be considered when assessing the reliability of a product. Since these will be unique for most products, every product demands a reliability program specifically structured to its circumstances. This is the basic concept in IEEE's standard P1332 [1].

The IEEE standard puts the responsibility on the supplier, working with the customer, to provide a product that satisfies the customer's requirements. The customer shall provide the supplier with an accurate and realistic description of the product requirements. From the customer's point of view, the outcome of this cooperation can be expressed as three questions [2]:

- Have I worked with the supplier to define and develop my requirements, and does the supplier understand them?

- Has the supplier developed a credible process to meet my requirements?
- Have I worked with my supplier to develop the metrics required to measure performance throughout product development to ensure that my requirements are met?

These questions translate into the three objectives that form the reliability program standard presented earlier, i.e., the supplier shall:

- Determine the customer's requirements and product needs.
- Meet the customer's requirements and product needs.
- Adequately verify that the customer's requirements and product needs are met.

The standard provides guidance to suppliers to plan a reliability program to fulfill these objectives that suits their design philosophy, the product concept, and the resources at their disposal. It stresses the importance of the freedom of a supplier to use innovative means to develop a product, i.e., neither the standard nor the customer shall specify the tasks to be performed. The standard deals with the activities that are required for development and production of reliable electronic equipment in a very general form.

The aim of the performance-based approach is to proactively incorporate quality into the design process by establishing a scientific basis for evaluating new materials, components, structures, and manufacturing technologies. A shift to a performance-based approach to assure quality of electronic hardware requires a well-structured, thought-out strategy where the activities, and the roles and responsibilities of everyone involved are clearly defined. A supporting infrastructure must be developed, and adequate resources must be allocated. This is in agreement with the requirements in ISO9000.

Since one of the basic ideas in the performance-based approach is that a supplier shall be free to determine how to assure the quality, there will be no standardized strategy specifying in details the activities that need to be done. It is up to every supplier to decide how he shall assure the quality of his products (as it is up to every customer to decide whether he approves the proposed way to assure the quality). Nevertheless, it is possible to list a general set of activities that ought to be incorporated in a performance-based approach for assuring the quality of electronic hardware. A proposed set of activities is given below.

The activities are as follows:

- Definition of product requirements and constraints during its expected design life.
- Definition of product life-cycle conditions and loading including both external and internal loading.
- Selection and characterization of alternative product architectures and manufacturing processes.
- Qualification of packaging concepts and manufacturing processes.
- Risk management and balance of functionality, quality, and cost requirements.
- Quality control and improvement of design, materials, parts, and manufacturing processes.
- Failure analysis and feedback of gained knowledge.

These activities are only to some extent a sequential flow of activities; iterations between the activities are necessary.

10.2 Activity 1: Product Requirements and Constraints

In this activity, the customer's requirements and any constraints shall be clarified. Customer requirements that need to be specified include performance, physical measures (size, weight, etc.), storage, transport, handling, maintenance, failure definition, reliability (acceptable failure rate), environmental compatibility, and life-cycle use. Constraints due to, for example, conflicts with legislation or the supplier's core competencies, culture, and goals must be defined.

The activity shall result in a requirements specification. The goal with the activity is to assure, both for the supplier and the customer, that the customer's requirements have been fully understood. Therefore, an effective dialogue between the supplier and the customer is required for ascertaining this has been achieved. The results of this process, the requirements specification, need to be approved by the customer. For consumer products, it may not be possible for the supplier to have this dialog with the customer. It will then be the responsibility of the marketing activities to capture the requirements (expectations) of the market.

10.3 Activity 2: Product Life-Cycle Conditions

Specification of product life-cycle conditions goes hand in hand with the product requirements specification activity, and these two activities form the first objective in IEEE P1332 [1]. The product life-cycle conditions are used for determining the loadings on the product during its life cycle, which are inputs for reliability assessment and development of design and manufacturing specifications, screens, and tests.

First, external loadings are defined, i.e., loadings due to the external environment. Loadings during the whole life cycle must be considered and characterized including loadings during manufacturing, testing, storing, transportation, maintenance, and use. Specific loading conditions may include:

- Temperature (steady-state, ranges, gradients, number of cycles)
- Humidity
- Contamination from production (flux residues, fingerprints, etc.) and use (corrosive gases, dust, etc.)
- Shock and vibration
- Pressure
- Radiation

- Power
- Current
- Voltage

The external loadings must then be converted to internal loadings, i.e., the actual loadings that materials, components, solder joints, etc., will be exposed to. The impact of power consumption and dissipation, internal radiation, shielding from external contamination and humidity, transformation of shock and vibration energy, aging, etc., must be considered.

Definition and characterization of external loadings need to be performed in cooperation with the customer. Since the adequacy of the reliability assessment is determined by the relevance of these input data, it is important that the data are correct. It is not enough to estimate (or guess) average values. Actual life-cycle conditions need to be specified. If these data are not available, it may be necessary to experimentally or through numerical simulation techniques determine the loadings. If it is still not possible to obtain credible data by these means, the worst-case design load must be estimated.

Transformation of external loadings to internal loadings must be done by the supplier. It is an iterative process executed during the whole design phase.

10.4 Activity 3: Selection and Characterization of Alternative Product Architectures and Manufacturing Processes

Design of electronic hardware involves selection of materials, components, interconnection techniques, interfaces, and manufacturing methods to realize the functionalities of the product. The objective is to use, as far as possible, well-known, proven technologies. However, the increased global competition forces companies to quickly adopt new technologies to avoid losing market shares. Choosing the right packaging concept may be crucial for the success of a product or even for the survival of the company. On the other hand, use of new immature technologies inevitably involves increased risk for manufacturability and reliability problems. A company must have the ability to accurately assess the risks associated with new technologies and determine when and how to use these technologies. With the ever-increasing number of available packaging concepts, this is becoming a true challenge for companies that will differentiate the winners from the losers.

The manufacturing of mobile phones is a good example showing how this development affects electronic companies. Manufacturers of mobile phones are exposed to a strong pressure to increase the functionality while at the same time decrease the size and weight. Even if it is not practical to decrease the size of a mobile phone any more, there is still a strong pressure to decrease the size of the circuitry as the mobile phones are integrated with other portable products. This is mainly possible by using new technologies. The fast development of new models of mobile phones with improved functionality forces companies to develop several new models each year. Choice of the wrong packaging solution will cause problems with manufacturability and reliability resulting in higher manufacturing costs and/or less reliable products compared to products manufactured by competitors who have chosen a better packaging concept. Furthermore, the short life cycle of mobile phones necessitates the time for development to be made as short as possible. Delays in the release of a new model due to manufacturability or reliability problems will likely cause decreased market shares and less or completely lost profit [3]. Therefore, materials, components, interconnection techniques, interfaces, and manufacturing methods finally selected during the design phase must have been sufficiently characterized in terms of manufacturability, testability, reliability, maintainability, environmental compatibility, etc. It must be known how the hardware performs over time when subjected to the specific manufacturing and application life profile conditions.

Owing to this development, many alternative packaging concepts have to be evaluated in parallel to find the best solution. Candidate materials, components, interfaces, designs, and manufacturing processes need to be characterized to enable an assessment of how they affect product quality. This involves specification of materials and components properties, buildup of components, printed boards, printed board assemblies, subsystems, complete system, and manufacturing processes. Results from this activity should be formulated in design and manufacturing specifications.

10.5 Activity 4: Qualification of Packaging Concepts and Manufacturing Processes

Qualification is usually defined as "The demonstration of the ability to meet all of the requirements specified for a product" [4]. Qualification testing is normally performed late in the development process of a product, often well after the design has been finished. In many cases, qualification testing is carried out on the complete product. Although this is in principle possible also in the performance-based approach as long as the requirements are relevant, in many cases it is not practical to demonstrate reliability at product level for hardware, for example fatigue life of solder joints. More importantly, this approach means that failure to pass the qualification testing may necessitate redesign of the product. A redesign is not only a costly process but it may also lead to lengthy delays in the release of the product and extremely high cost due to lost market shares.

To avoid redesigns, qualification testing of hardware should be done during the initial product development. This can be achieved by demonstrating the ability to meet specified requirements by showing that the design and manufacturing processes are under such control that specified requirements will be met. That is, it is the processes rather than the end product that are qualified.

In a wide sense, qualification of a product then includes all activities required for assuring that the customer requirements are met, i.e., Activities 1–7 discussed in this chapter. The objective of the first three activities is to transform the customer's requirements into adequate design and manufacturing specifications for the product. The objective of this fourth activity is to assess how the alternative packaging concepts, designs, and manufacturing processes affect manufacturability, reliability, maintainability, environmental compatibility, etc.

10.5.1 Manufacturability

It is important that the hardware design team understand materials constraints, available processes, and manufacturing process capabilities if they shall be able to select materials and parts, and construct architectures that promote manufacturability, reduce the occurrence of defects, and increase yield and quality.

The choice of materials, parts, and interconnecting techniques must be compatible with the assembly equipment and processes that will be used. For example, a certain free space is required around components to allow placement and rework of components. The difficulties to rework solder joints of area array components requiring that the components be removed necessitate special considerations. The type of equipment used for rework will determine the free space required around the components to facilitate the replacement of the components and prevent that the reliability of adjacent components are affected by the rework process. It may also be necessary to limit the types of components that can be mounted on the opposite side of the board.

Another important issue is routing. For example, area array components must be routed out from under the component. The number of I/Os, pitch, line widths, insulation distances, etc., will determine the board area and the number of board layers required for the routing. Since solder joints to area array components are not accessible for testing pins, this will affect the testability of the printed board assembly. If testability is important, routing can be done to special test pads around the component or on the opposite side of the board. However, this will likely increase the board area required for the component.

To facilitate the design work, design rules related to the equipment and processes used should be developed.

10.5.2 Reliability

The process for qualification of packaging concepts and manufacturing processes involves a quantitative estimate of the failure rate or probability of failure. Possible failure modes and associated failure mechanisms for the product can be determined with knowledge of life-cycle loading conditions, product architecture, and manufacturing processes. The capability of the product to perform reliably shall be assessed over its entire intended life-cycle environment, based on the local environments. Upper and lower operating limits need to be determined.

All possible failure mechanisms must be considered to assure the reliability of the product. Knowledge of the physics-of-failures is important for construction of relevant tests and screens for assurance of the reliability. All critical design, manufacturing, and operating parameters that will affect the failure mechanisms must be defined. Also, the acceptable range of variability of these parameters must be determined.

10.5.2.1 Assessment of Failure Probability

The probability for failure due to various failure mechanisms can either be assessed from previous experience or, when experience is not available, from reliability tests. If it can be assessed from previous experience, large cost will be saved, but that is usually only possible to a limited extent, especially when new technology is used. The types of tests to perform depend on the type of failure mechanisms. Reliability tests are usually associated with wearout failure mechanisms and, therefore, tests for wearout failures will be discussed first. However, for most products, failures due to defects and overstress are far more frequent. Hence, it will be necessary to pay more attention to tests for failures due to defects and overstress in the future.

Wearout failures. For wearout mechanisms, accelerated reliability tests are usually required to achieve test results in reasonable time. Acceleration of a test can be achieved in two ways, which may be combined [5]. The frequency of the occurrence that causes failure can be accelerated or the severity of the conditions causing the failure can be increased. Generally, a failure mechanism that is a continual process going on most of the time, for example a corrosion process, can best be accelerated by increasing the severity of the conditions that cause failure, whereas a failure mechanism that is caused by a number of sequential events of rather short duration best can be accelerated by increasing the frequency of the events.

Accelerated tests require careful planning if they are to represent the actual usage environment and operating conditions without introducing extraneous or nonrepresentative failure mechanisms. Failure mechanisms that dominate under normal usage may lose their dominance as the stress is elevated, whereas failure mechanisms that are dormant under normal usage may contribute to failures or even become the dominant failure mechanisms at high stress levels. Obviously, the risk for accelerating wrong failure mechanisms is largest when acceleration is achieved through increasing the severity of the conditions causing the failures. Nevertheless, it can also happen when acceleration is achieved through increasing the frequency of the occurrence that causes failure.

The fatigue of solder joints due to repeated changes in temperature that has been discussed in Chap. 5 is taken as an example. The fatigue life is normally evaluated using thermal cycling tests. Acceleration is achieved both through increasing the severity (larger temperature range) and the frequency of occurrence (temperature cycles per time unit). Thermal cycling is often performed between −40 and +125°C or even between −55 and +125°C to achieve a high acceleration factor. However, it is recommended that thermal cycling should not be performed outside the range 0 to +100°C unless the products will be exposed to temperatures below 0°C and/or above +100°C [6]. The reason is that in the temperature region from about −20 to +20°C, a primarily stress-driven solder response to applied loads at lower temperatures change to a primarily creep/stress relaxation response at higher temperatures. Thus, the damage mechanism will be different if thermal cycling is performed down to −40 or −55°C compared to cycling down to 0°C. Furthermore, if the selected high temperature extreme comes close to the T_g (glass transition temperature) for the laminate material in the board substrate, this may have a large impact on the failure mechanism [5]. If low T_g FR-4 is used, which is the most common board laminate, it has a T_g of about 135°C but the laminate starts to become soft at somewhat lower temperatures, which will cause a decrease of the stress applied to the solder joint. Therefore, the high temperature extreme should not be higher than +100°C when the laminate is low T_g FR-4. If laminates with higher T_g is used, the high temperature extreme can be higher. If the actual product will be exposed to temperatures below 0°C and/or above +100°C, it is recommended to add a number of cycles similar in nature and number to actual use.

This means that for some applications that are exposed to large temperature changes, for example some under-the-hood applications, the acceleration factor cannot be increased much by increasing the severity, in this case the temperature range. Hence, in such cases, increasing the acceleration factor must be achieved mainly through increasing the frequency of the temperature cycles. The frequency is determined by the cycle time. Since a thermal cycle consists of temperature ramps (up and down) and dwell times at the temperature extremes, the ramps should be as fast as possible and the dwell times as short as possible to get an acceleration factor as high as possible. The most extreme acceleration is achieved if the test vehicle is dipped alternately in two liquids of different temperatures. However, it has been found that this may produce misleading results. Rapid temperature changes cause large transient thermal gradients resulting in warpage of both components and board substrate. The warpage will cause both tensile and shear stresses where the tensile loading dominates. Even assemblies with matched CTEs will exhibit solder joint failures when subjected to thermal shock. In contrast, slow thermal cycling results mainly in shear loads and the eventual failure occurs from an interaction of shear fatigue and stress relaxation. Consequently, tests involving rapid temperature changes testing for purposes of evaluating solder joint reliability are only appropriate if rapid temperature changes at board level are indeed a field condition encountered by the product, which is a very unusual condition. In normal cases, the temperature ramp should be less than 20°C/min [6].

Then remains decreasing the dwell time as a means to decrease the cycle time. Because fatigue of solder joints is mainly due to creep, a certain dwell time is required to allow for stress relaxation. If the dwell times are too short, the number of cycles required to produce a failure will in fact be increased. Fifteen minutes dwell time at each temperature extreme is recommended in IPC-SM-785 for lead-based solders. Longer dwell times may be needed for lead-free solders since creep is usually much slower in these.

This example shows how important it is to have knowledge of the physics-of failure to design adequate accelerated tests. All parameters affecting the failure mechanism must be understood, and stress levels must be optimized to accelerate the relevant failure mechanisms but not nonrelevant. Not only operating (testing) parameters affect the test results but also design and manufacturing parameters and these must also be defined. To stay with fatigue of solder joints as an example, the geometry of the solder pads on the component and on the printed board will affect the fatigue life [7, 8]. As an example, for a BGA with high-temperature melting balls, the solder paste volume printed on the solder pads is critical for the fatigue life [9]. Thus, reliability testing must also include evaluation and determination of the acceptable ranges of variability (process windows) of these and many other parameters to assure the reliability of the end product. Simulation tools can be valuable in the process of identifying the parameters that will be important for various failure mechanisms and for optimizing material properties and process parameters. Simulation is normally much faster and less expensive than accelerated tests and can sometimes complement or even replace them.

Failures are in many cases due to a combination of failure mechanisms. The ideal solution is to find a test that accelerates all failure mechanisms simultaneously in the same manner that will occur during a product's use, but that is rarely achievable. Tailoring a program of consecutive tests is usually the only solution. Interactions between various failure mechanisms must then be considered so that the order of performance of the tests gives the right types of interactions. As an example, vibration may interact with thermal cycling, leading to shorter fatigue life of solder joints. In some test chambers, vibration testing can be performed simultaneously with thermal cycling. If such a chamber is not available for testing, a consecutive test must be performed. The order in which the tests are performed can be expected to affect the test results since vibration likely has much more impact on crack propagation than on crack initiation. A possibility is to expose the test vehicles alternating to thermal cycling and vibration.

The goal of accelerated tests is to estimate the failure rates of wearout mechanisms during a product's life. Hence, it must be possible to quantitatively extrapolate from the accelerated conditions to the usage conditions with some reasonable degree of assurance. That is, the acceleration factors for the tests must be determined if they are not already known.

Overstress failures. The assessment of the probabilities for overstress failures requires a different strategy. If the failure mechanism is a true overstress mechanism, the goal is to find the stress level at which failures occur. This can be achieved by successively increasing the stress load (step stressing) until a failure is observed. The probability of failure can then be assessed by estimating the likelihood that the determined critical stress level will be exceeded in usage.

In many cases, failure mechanisms are due to a combination of wearout and overstress, such as electrochemical migration. A combination of accelerated testing and step stressing can then be a useful approach [10].

As for wearout mechanisms, the design and manufacturing parameters that affect the failure mechanism must be defined, and the acceptable ranges of variability of these must be determined. Overstress failures can also be affected by interactions between various types of stresses. Printed board assemblies that may be exposed to condensation of water are usually conformally coated to prevent electrochemical migration. Exposures to low temperatures, fast temperature changes, or vibration may cause cracking of the conformal coating exposing biased surfaces. This may ruin the protection against electrochemical migration. This must be considered when determining the tests to be performed.

Thus, the process of minimizing the risk for overstress failure is better described as design-for-robustness than design-for-reliability.

Failures due to defects. Assessment of the probabilities for failures due to defects require yet another approach. Since defects are product deficiencies that should not be present, assessment of failure probability consists of two parts. First, the probability that a defect shall be present must be assessed. Then, the probability that the defect will cause a failure must be estimated.

The probability for that various defects shall be present can to some extent be estimated through records of previous production, for example solder defects. However, the true nature of defects, i.e., deficiencies that is unintentional and caused by chance, makes it difficult to foresee all types of possible defects. This necessitates that quality controls are carried out during the development and manufacturing of a new product to detect any defects that may occur. The design of tests for quality controls is discussed in Sect. 10.7.

In order to assess the probability that a defect will cause a failure, the magnitude of the defect must be known. Often the magnitudes of defects vary within a certain range. Misalignments of components and meager solder joints due to insufficient print of solder paste are examples of such defects. This range and the distribution of the variation within this range must be known for an adequate assessment of the probability of failures. Again, experience from previous production is the best source for this information.

Execution of reliability tests. Since the objective with the assessment of failure probability is to qualify processes rather than the end product, reliability testing and ascertaining should be performed at every stage in the product development process. The reliability of materials, components, and printed boards that are bought from suppliers must, as far as possible, be ascertained by these suppliers. Of course, they must also use a performance-based approach for quality assurance. The manufacturer of the end product then becomes buyer and must fulfill the responsibilities of a buyer discussed in Activities 1 and 2.

To facilitate production of silicon devices according to the performance-based approach, JEDEC has developed a standard, or rather guideline, JESD34, "Failure-Mechanism-Driven Reliability Qualification of Silicon Devices" [11]. In principle, it covers the same area as IEEE P1332, but it is more detailed and has a specific focus on production of silicon devices. It is pointed out in JESD34 that for the standard to reach its full effectiveness, the original equipment manufacturer (OEM) and the supplier (device manufacturer) must develop partnerships. The OEM should accept the reliability qualification process performed by the supplier, per the guidelines of the standard, in lieu of the specific (special) reliability qualification process that many OEMs require today. Furthermore, since component manufacturers lack much of the knowledge needed for tailoring a relevant test program, it is pointed out in JESD34 that OEMs must be committed to the collection and analysis of field data and be prepared to share applicable data with component suppliers. It will then be the responsibility of the component manufacturer to identify all potential physical failure mechanisms and evaluate the impact on reliability. Potential failure mechanisms must include mechanisms that may be introduced by subsequent levels of manufacturing consistent with the intended use of the product and mechanisms that may occur over the expected lifetime of the product, in the intended field application conditions to which the product is expected to be subjected. It will be the OEM's responsibility to compare the manufacturer's application assumptions against the intended application conditions. The JESD34 standard was rescinded in 2004 and instead replaced with JESD94 [12], which also has a performance-based approach. Today, many component manufacturers work along these lines [13].

There are also other reasons why reliability testing should be performed as early and at as low a product level as possible. The earlier a reliability problem can be detected, the less expensive will it be to correct it. Furthermore, some reliability tests may take quite long time to perform. Adequate evaluation of the fatigue life of solder joints for applications where longevity is required may take up to 1 year or even more for certain applications where very high reliability is required [6]. It will save a lot of time if the component manufacturer has ascertained the fatigue life of solder joints to the component.

Test vehicles. As discussed in the previous section, many of the reliability tests can and should be performed well before prototypes are available. This means that special test vehicles often must be used for the evaluations. The test vehicles must be representative of materials, printed board substrate, and production processes that will be used for producing the end product, in some cases also of normal changes due to aging, to adequately reflect true conditions. To stick to fatigue of solder joints as an example, some of the more important factors that will affect the fatigue life are properties of the printed board and solder, solder pad geometry and plating, soldering process, and solder grain coarsening due to aging [6]. Consequently, a manufacturer of components will not be able to assess the fatigue life of solder joints for all conceivable applications. He can only do it for some typical applications. A manufacturer of printed board assemblies may need to complement the evaluation done by the component manufacturer.

Furthermore, some failure mechanisms are difficult to evaluate by testing the end product or prototypes. It may then be better to use special test vehicles. For example, to evaluate the fatigue life of solder joints, it must be possible to register open circuits in any of the solder joints during a thermal test. This is best achieved by using a special test board and test components with daisy-chain interconnection.

However, the assessment of the probability of some failure mechanisms may need to be verified or are best carried out on the end product or a prototype. Examples of such failure mechanisms are Electromagnetic Interference (EMI) and those caused by dropping the product to the floor, i.e., overstress failure mechanisms. It may also be necessary to verify that the local loads are consistent with those that have been assumed during the assessment of the probability or failure rates of the various failure mechanisms, for example the temperatures of components.

Failure analysis. Reliability testing is of little value if it is not followed by adequate failure analysis where the root causes of failures are determined. It is necessary to know the root causes to determine which measures should be taken to improve the reliability. It is also necessary for assessment of the relevance of the reliability tests.

10.5.3 Maintainability

Most electronic products need to be maintained or repaired during usage. For some products, the hardware may need to be updated. The design will have a large impact on the maintainability. It will affect both the availability of the product (how long it will take to maintain or repair) and the lifetime cost of the product. Therefore, it is important that maintainability issues are covered during the design of a product.

10.5.4 Environmental Compatibility

During the 1990s, increased attention has been put on environmental compatibility. The use of CFCs (chlorinated and brominated solvents) as cleaning solvents has been banned. In the European Market, there is a proposal for prohibiting the use of lead and some brominated flame retardants in the production of electronics. Not only legislation may force companies to convert to materials and processes that are more environmentally friendly. The increased environmental awareness among customers is in many cases a stronger driving force for companies to adopt a profile of a company that cares for the environment. For cost-effectiveness, environmental compatibility issues must be covered already during the design phase of a product.

10.6 Activity 5: Risk Management and Balance of Functionality, Quality, and Cost Requirements

When selecting the final packaging concept and manufacturing processes, trade-off between various requirements (functionality, manufacturability, reliability, cost, etc.) is necessary for finding the optimal solution for a specific application. For space applications, manufacturability and cost issues often are less important than reliability issues, whereas for low-performance consumer products, it is usually the opposite.

Risk management is an important part of the balancing of various requirements. It can be split into risk management of parts that are bought from suppliers and risk management of the actual companies' activities.

10.6.1 Risk Management of Supplied Materials and Parts

The adequacy of qualifications done by suppliers must be scrutinized. If a qualification is found not to be adequate, what it will cost and how long it will take to complement the supplier's qualification must be assessed. Availability and suppliers' ability to produce materials and parts with consistent quality must be gauged. When new technology is used, a higher infant mortality and larger variation in quality can be expected. A supplier's policy to notify about changes in materials and manufacturing processes is important since it may affect reliability and variability in a part's characteristics although the functionality may be the same.

Owing to the short life cycle of many components, there is a risk that some components may become obsolete during the time span a product is developed and used. The risk for components becoming obsolete must be carefully considered and assessed to aid in the selection of these. Proactive plans to handle components becoming obsolete including a second source may mitigate problems due to obsolete components.

10.6.2 Risk Management of Manufacturing Processes and New Technologies

The capabilities of the manufacturing processes to deliver products fulfilling the quality requirements and the associated cost need to be assessed for the various packaging concepts and manufacturing processes. Calculation of the manufacturing cost requires that the yield and cost for rework and eventual cassation are estimated. If packaging concepts and manufacturing processes used are new, either to the company or to everyone, it may be difficult to estimate these figures. Hence, new packaging concepts and manufacturing methods are associated with a higher risk. To decrease this risk, it may be advisable to test them prior to using them for a product to facilitate better assessment of the cost associated with them.

New technology may also require investment in new equipment, which will add to the cost. If untested bare dies are used, cost for testing must be considered including prediction of percentage defective dies.

10.6.3 Failure Modes and Effects Analysis

Failure modes and effects analysis (FMEA) is a useful evaluation tool for comparing alternative packaging solutions from a reliability point of view. The objective with FMEA is to analyze and assess the effects of potential failures in a product. Failure effects may be considered at subsystem or overall system levels. The possible failure modes are identified for every part in the product, and the effects on product operation and personnel safety are analyzed. A fish-bone diagram of the product, showing all the possible ways in which the product can be expected to fail, is often done in this process. FMEA can be started as soon as initial design information is available, and then proceed iteratively as the design evolves.

10.6.4 Protective Measures

Various measures can be taken to improve the reliability, from low cost with small impact on reliability to more expensive with larger impact on reliability. The use of fuses is an old and well-known method to prevent failures or limit the effects of a failure. This method can be extended by the use of various types of sensors that shut down the equipment in cases of risk for failures. It can be temperature, humidity, or more sophisticated sensors. It is a rather cheap method to prevent failures, but it has the drawback of causing interruptions in the use of the equipment.

Another approach is to lower the loading of the parts that may fail. If the failure is caused by high temperature, this can be achieved by improving the cooling in various ways. In humid and corrosive environments, conformal coating, encapsulation, or use of airtight cover can improve reliability.

For equipment with very high reliability and availability requirements, redundant systems may be the best solution, but this comes at a cost.

Each method has its advantages and drawbacks. The cost for the extra safety has to be measured against the worth of the improved reliability.

10.7 Activity 6: Quality Controls and Improvement of Design, Materials, Parts, and Manufacturing Processes

Quality controls create cost without adding value to the product and should therefore be kept at a minimum. Ideally, if design and manufacturing processes were under complete control, it would not be necessary to do any quality controls at all. This is rarely the case. The larger the uncertainties are about the outcome of design and manufacturing processes, the larger is the need of quality controls. New immature technologies require more quality controls than mature technologies.

This cost, including estimated cost for repair and cassation, must also be considered when balancing the cost for the alternative packaging concepts and manufacturing processes.

The purpose of quality controls is to detect defects in the design, materials, parts, and manufacturing processes. A defect is then defined as a design, material, part, or manufacturing process attribute that is outside acceptable ranges and thus has the potential to compromise product reliability. Selection of a component that will have solder joints with nonacceptable fatigue life can then be defined as a design defect.

Quality controls should be planned with the purpose of detecting defects as early as possible in the development and manufacturing of a product. The earlier a defect is detected, the less expensive will it be to take necessary corrective actions. More importantly, early corrections of weaknesses in design and manufacturing processes will lead to a faster release of a more mature and robust product.

10.7.1 Design Defects

In a wider sense, the qualification of packaging concepts and manufacturing processes discussed in Sect. 10.5 is a quality control of the design. Therefore, this section will deal with how to detect weaknesses in design, including choice of materials and parts, that have been overlooked in the qualification process and then mainly weaknesses that will cause overstress failures. Wear-out failure mechanisms are more limited in numbers and therefore easier to foresee and thus not as likely to be missed. Also, it is much more difficult to develop a quality control tool for detecting overlooked wear-out failure mechanisms.

A tool developed for detecting weaknesses in a design is Accelerated Stress Testing (AST). The purpose with AST is to apply high levels of stress to quickly precipitate weaknesses to failures. By analyzing these failures, the weakest points in the design will be identified. Thereby, measures can be taken to correct the weaknesses and improve the robustness of the product. Since the applied stress levels usually are higher than those that the product will be exposed to during normal usage, failures may be caused by inadequate failure mechanisms. Thus, the relevance of occurred failures must always be evaluated.

AST is often confused with Accelerated Life Testing (ALT). In ALT, the tests are designed to simulate the product's life to assess its expected lifetime, whereas in AST, the tests are designed to stimulate failures that might occur during the product's life. The goal of using ALT on a product is usually that the product shall pass preestablished stress levels without any failure whereas AST has no such preestablished levels. The goal with AST is that failures should occur to identify possible failure modes and thereby find out potential weaknesses in the product.

Step stressing is usually utilized during AST. That is, the test is started with a rather low stress level. The stress level is then increased in steps until a failure occurs. When a failure has occurred, the stress level may be reduced to analyze whether it is a soft failure (a failure where the product resumes operation when the stress level is reduced) or a hard error (when the product does not resume operation). If possible, the failure is repaired so that the test can be continued at higher stress levels to find more failure modes. Since the purpose of AST is to stimulate weaknesses or defects to failures so that they can be detected, it is not necessary that the failure mechanism during AST is identical with the one that might take place in field conditions. However, the higher the stress level is increased above the operation level, the larger is the risk for failures that would never occur in field conditions, for example melting of materials. Therefore, after AST has been finished, the root causes of the detected failures must be determined so that the relevance of the failures can be assessed.

The goal is to precipitate as many as possible of all latent weaknesses that might cause failures during a product's lifetime. Therefore, as many relevant types of stresses as possible should be used. Examples of stresses that could be applied are low and high temperatures, temperature cycling, humidity, vibration, shock, voltage, power, and various forms of radiation. If possible, stresses should be combined. A common combination is temperature cycling and vibration.

Besides identification of possible failure modes, AST will give information about operational and destruct limits of the product. This information is useful for checking that the risks for overstress failures have been correctly assessed when balancing the various requirements.

Usually, it is easy to design tests that will precipitate weaknesses to failures, but if the failures are not detected the test has accomplished nothing. Thus, it is of paramount importance that all possible types of failure modes can be detected during the test, including soft and intermittent failures. This requires continual monitoring of the functionalities of the product. For this reason, and because AST should be performed as early as possible in the development process, it is usually performed as soon as the first functioning prototypes have been produced. However, it can and should be used at all levels deemed necessary (component, printed board assembly, subsystem, and complete system). The earlier a flaw can be detected, the less costly will it be to correct it.

In the 1990s, a form of AST known as HALT has gained increased popularity, especially in the USA. HALT is the acronym for Highly Accelerated Life Testing.

Unfortunately, this is a misleading name, since it is a true stress test and not a life test, which causes confusion and misunderstandings when people first get into contact with it. HALT is perhaps best described as a package of stress tests consisting of exposure to low and high temperatures, temperature cycling, multiple axis vibration, and combination of temperature cycling and vibration. The purpose with HALT is to find design flaws.

10.7.2 Defects Caused by Manufacturing Processes

Traditional practice relies heavily on inspection as the main method to assure a defect-free product. Inspection is usually done after the various manufacturing processes and to a large extent by visual inspection, either with or without magnification. This is a costly process, partly because visual inspection is a labor-intensive process and partly because defects require costly repair work. The approach in performance-based quality management is to proactively prevent defects from occurring by effective control off all process parameters that may cause failures. This is achieved through statistical process control (SPC) and continual process improvement. SPC has a long history that goes back to original work done by Shewhart in the 1920s [14]. It forms the basis in Total Quality Management (TQM) developed by Deming and others that was so successfully adopted by Japanese companies in the 1950s.

During the qualification process of packaging concepts and manufacturing processes, all critical manufacturing parameters and the acceptable range of variability of these shall have been determined. SPC is used to control that the variability of the parameters are hold within the acceptable ranges. This involves measuring of the process parameters, a sort of inspection, but in contrast to the traditional practice, it is done proactively and to check the outcome of the process and not the outcome on the product. As an example, insufficient print of solder paste may result in meager solder joints with decreased fatigue life or even in open solder joints. In the traditional practice, this is dealt with by visual inspection of all solder joints after the soldering operation. In performance-based quality management, it is dealt with by controlling the volume of solder paste printed prior to soldering. If the solder paste volume is insufficient on any solder pad, the process defect can be corrected before it results in a product defect.

If the acceptable range of variability is not known for all process parameters, SPC can be used to get them under control. Alternatively, SPC can be used to improve the quality by identifying and narrowing the process windows of the process parameters that have largest impact on quality. The outcome of the processes on product quality is then determined as a function of the process parameters. Of course, this requires some measuring or inspection of the product until all process parameters are under control or have been optimized.

Ideally, SPC should be enough to ensure that manufacturing processes do not introduce defects in the product. However, in practice that is only possible for very mature processes and even then this is not always feasible. Hence, there is a need for methods designed to detect process-related defects. Furthermore, it is not always easy to know what constitutes a process-induced defect. As for design defects, AST can be used to precipitate manufacturing defects to failures and thereby make them more easy to detect. It is then often called Environmental Stress Screening. If HALT has been used to characterize the design of the product and remove design flaws, the operational and destruct limits are known for the product. This is used in HASS (Highly Accelerated Stress Screens), also developed by Gregg Hobbs [15], to create a very highly accelerated test that will quickly precipitate defects introduced by manufacturing processes into failures. HASS starts with a precipitation screen. The product is then exposed to a combination of thermal cycling and vibration at stress levels below the levels that will cause hard failures. If the precipitation screen causes soft failures, it is followed by detection screen where the stress levels are decreased to levels at which no soft failures should occur in a defect-free product. As for HALT, continuous monitoring should be performed during the whole test to detect soft, hard, and intermittent failures.

HASS can be used both for checking that the manufacturing processes are under control when production is started up for a new product and for checking that the quality does not deteriorate with time. The latter can, for example, be due to a parameter drift in equipment or that a supplier delivers a bad batch of components. Since HASS utilizes very high stress levels, it may take some of the products life, even though it passes the test without failure. Therefore, if products that have been exposed to HASS will be delivered to customers, it must be verified that the test does not take too much of the products life. A rough method to do that is to expose the product to repeated HASS, for example 10 or even 100. If a defect-free product passes the prescribed numbers of HASS without failure, the HASS is approved (proof-of-screen).

10.8 Activity 7: Failure Analysis and Feedback of Gained Knowledge

The main concept of TQM is continual improvement of quality and productivity. This is, of course, still a useful concept. Deming put the main effort on improvement of the manufacturing processes after the design had been finished and the production had been started up for a product. In performance-based quality management, the main focus has shifted from the manufacturing phase to the design phase. Continual improvement is still an important part, i.e., the seven activities described in this chapter should be continuously improved. Thereby, the ability to produce quality products will be improved and less work and time will be required to develop the next product.

The best source for improving the qualification process of a new design is the experience gained during design, manufacture, and use of previous products. Therefore, routines must be established that assure that knowledge gained is fed back and used for improving the seven activities discussed in this chapter. Failures that occur during testing and use must be analyzed and the root cause (physics-of-failure) must be determined.

Exercises

10.1 Why is the traditional standards-based approach to assure reliability not always relevant to modern microsystems?

10.2 Describe the three objectives in IEEE's standard P1332, Standard Reliability Program for the Development and Production of Electronic Systems and Equipment.

10.3 Give examples of loadings during a product's life cycle that may affect reliability.
10.4 How should tests be designed to assess the risks for early failures during a product's life?
10.5 How should tests be designed to assess the risks for failures due to aging and wearout?

References

1. IEEE P1332, Standard Reliability Program for the Development and Production of Electronic Systems and Equipment, IEEE, 1998.
2. A. Malhotra, A. Strange, L. Condra, I. Knowles, T. Stadterman, J. Boivin, A. Walton and M. Jackson, "Framework for an Objective and Process Based Reliability Program Standard", Communications in RMSL, 1996, 25–30.
3. B. McDermott and M. Peterson, "Concurrent Engineering for the New Millennium", HDI, 4 (7), 2001, 20–24.
4. IPC-T-50F, Terms and Definitions for Interconnecting and Packaging Electronic Circuits, IPC, 1996.
5. J. Glazer, Reliability of printed circuit assemblies, in C.F. Coombs (Ed.), Printed Circuit Handbook, 4th Edition, McGraw-Hill, New York, 1995, Chapter 37.
6. IPC-SM-785, Guidelines for Accelerated Reliability Testing of Surface Mount Solder Attachments, IPC, 1992.
7. A. Mawer, D. Cho and R. Darveaux, "The Effect of PBGA Solder Pad Geometry on Solder Joint Reliability", Proc. Surface Mount International, 1996, 127–135.
8. R. Rörgren, P. E. Tegehall and Carlsson, "Test Methods and Reliability Evaluations of BGA Packages for Automotive Electronics", Proc. ISHM Nordic, 1997.
9. P. E. Tegehall and B. Dunn, "Assessment of the Reliability of Solder Joints to Ball and Column Grid Array Packages for Space Applications", ESA STM-266, ESA Publications Division, Noordwijk, 2001.
10. P. E. Tegehall, "Reliability Verification of Printed Board Assemblies: A Critical Review of Test Methods and Future Test Strategy", Proc. Surface Mount International, 1998, 359–382.
11. JESD 34, Failure-Mechanism-Driven Reliability Qualification of Silicon Devices, JEDEC Solid State Technology Association, 1993.
12. JESD 94A, Application Specific Qualification Using Knowledge Based Test Methodology, JEDEC Solid State Technology Association, 2008.
13. F. Wulfert and H. Tiemeyer, "Qualification for Reliability in Time-to-Market Driven Product Creation Processes, Tutorial 2", Proc. 11th European Symposium on Reliability Of Electronic Devices, Failure Physics and Analysis, Dresden, 2000.
14. N. C. Noel, The Four Pillars of Wisdom: A System for 21st Century Management, the Swiss Deming Institute, 2000, http://www.deming.ch.
15. G.K. Hobbs (Ed.), Accelerated Reliability Engineering: HALT and HASS, John Wiley & Sons, Inc., New York, 2000.

Chapter 11
Experimental Tools for Reliability Analysis

Abstract In this chapter, several basic types of experimental tools for different situations of reliability analysis are introduced, together with the working principles. After that, tools being used to do accelerate testing are also presented.

Optical microscopy (OM), scanning electron microscopy (SEM), energy-dispersive X-ray (EDX), scanning acoustic microscopy (SAM), and moiré interferometry are used to measure the structure and geometry of the testing sample. Besides, low-cycle fatigue, shear, humidity, temperature, thermal shock, and thermal cycling tests could be done with the help of special types of machines.

11.1 Optical Microscopy

For optical microscopy (OM) observations, a microscope with magnification lens from 100 to 1,000× is used. Optical microscopy is used mainly to measure cracks inside the solder joints, and equivalent functions for ECAs.

11.2 Scanning Electron Microscopy

Scanning electron microscopy (SEM) provides a unique tool for microstructural studies. It produces images of solid material surfaces such as optical microscopy; however, its advantage compared to optical microscopy is its large depth of field over a wide range of magnifications, which makes SEM one of the most extensively used instruments in this research area today.

SEM uses electrons instead of light to form an image. A beam of electrons is produced at the top of the microscope by heating of a metallic filament. The electron beam follows a vertical path through the column of the microscope. It makes its way through electromagnetic lenses, which focus and direct the beam down toward the sample. Once it hits the sample, other electrons are ejected from the sample; however, not all of them are detected and used for information. Detectors collect the secondary or backscattered electrons and convert them to a signal that is sent to a viewing screen, which produces an image. This processing has to be done in vacuum; otherwise, the

J. Liu et al., *Reliability of Microtechnology: Interconnects, Devices and Systems*, DOI 10.1007/978-1-4419-5760-3_11, © Springer Science+Business Media, LLC 2011 transmission of the beam through the electron optic column could be scattered by the presence of other molecules, which can come either from the sample or the microscope itself, and in the end obscure details in the image.

Materials to be analyzed with SEM have to be conductive; otherwise, the image will be blurred. A sputter coater makes nonconductive materials conductive by producing a thin gold coating on the surface of the sample. Otherwise, there are no limits to SEM use.

An SEM analysis presents the advantage of giving a combination of both 2D-imaging, where it is possible to distinguish the different depth range by the intensity of the backscattered and secondary electrons, and chemical analysis, by means of energy-dispersive X-ray (EDX) analysis.

11.3 Energy-Dispersive X-Ray

EDX is a common accessory that gives an SEM a very valuable capability for elemental analysis by measuring the energy or wavelength and intensity distribution of X-ray signals generated by the focused electron beam on the specimen. When the incident beam bounces through the sample creating secondary electrons, it leaves thousands of the sample atoms with holes in the electron shells where the secondary electrons used to be.

If these "holes" are in inner shells, the atoms are not in a stable state. To stabilize the atoms, electrons from outer shells drop into the inner shells; however, because the outer shells are at a higher energy state, to do this, the atom must lose some energy. It does this in the form of X-rays.

This tool allows simultaneous nondestructive elemental analysis of the sample. The X-rays are emitted from a depth equivalent to how deep the secondary electrons are formed. Depending on the sample density and accelerating voltage of the incident beam, this is usually from 1/2 to 2 μm in depth; therefore, EDX is not a surface technique.

Elemental mapping by means of EDX. It is also possible to map the elements found in an SEM image by X-ray analysis. By setting windows around the peaks of specific elements, the SEM software can scan the sample and create digital images or maps of each element. By placing dots on the screen when an X-ray count of the particular element is received, an image is formed that mimics the SEM image, except the contrast is formed by the elemental X-ray emission.

11.4 Scanning Acoustic Microscopy

Scanning acoustic microscopy (SAM) uses acoustic impedance to produce high-resolution images of the interior structure of a sample to study surface and subsurface features and also detect "difficult-to-find" defects, such as interfacial separation/delamination, solder-ball delamination, and die attach voiding.

A scanning acoustic microscope works on the principle of propagation and reflection of acoustic waves (with high frequency) at interfaces where a change of acoustic impedance (AI $=$ density $\times$ velocity) occurs. The sound wave is propagated through water into the sample. At positions where an impedance (velocity $\times$ density) change occurs, the sound waves are reflected back. The distribution of these reflected waves is mapped and shows how impedance changes within the sample.

Since the working frequency of the acoustic waves can be varied, the depth of penetration of the acoustic waves into the sample and the resolution of microstructural features also vary correspondingly (the higher the frequency, the higher the resolution and the lower the penetration). The lens is scanned in a raster pattern over the specimen to form an image.

In this thesis, the equipment used for the SAM analysis was a Sonoscan D6000 with a pulse-echo acquisition mode (reflective mode) and a transducer frequency of 30 MHz.

11.5 X-Ray

As the wavelengths of light decrease, they increase in energy. X-rays are electromagnetic radiations of wavelength about 1 Å (10^{-10} m), which is about the same size as an atom, and have high energy. Different materials have different densities and pass different amounts of radiation. Dense materials absorb more X-rays and are therefore seen as dark shadows. Lighter materials let pass more X-rays and are therefore lighter in contrast.

The principle of an X-ray machine is quite simple. Within the machine, there is an X-ray tube with an electron gun inside that shoots high-energy electrons at a target made of heavy atoms, such as tungsten. X-rays are emitted because of an atomic de-excitation process induced by the high-energy electrons that were shot at the target. There are two different atomic processes that can produce X-ray photons. One is called Bremsstrahlung where the electrons slow down after swinging around the nucleus of a tungsten atom and lose energy by radiating X-rays. A lot of photons are in reality produced, but none of the photons has more energy than the electron had to begin with. After emitting the spectrum of X-ray radiation, the original electron is slowed down or stopped.

The other process is called K-shell emission, where the incoming electron from the electron gun gives enough energy to knock a K-shell electron in the tungsten target atoms, and a tungsten electron of higher energy (from an outer shell) can fall into the K-shell. The energy lost by the falling electron shows up in an emitted X-ray photon. Meanwhile, higher-energy electrons fall into the vacated energy state in the outer shell, and so on. K-shell emission produces higher-intensity X-rays than Bremsstrahlung and emits X-ray photons at a single wavelength. Both ways involve a change in the state of the electrons.

Fig. 11.1 LCF Microtesting System Model 228

11.6 Low-Cycle Fatigue Testing

Low-cycle fatigue (LCF) is regarded as one of the main mechanisms of failure of solder joints. Generally, the amplitude of plastic strain of LCF is relatively larger, so the fatigue life is usually less than 10^4 cycles in the low-cycle region.

An isothermal mechanical fatigue tester is often employed as an effective device to evaluate the LCF behavior of solder. During isothermal LCF testing, samples are usually loaded at a constant cycle stress or strain at constant temperature, normally at room temperature.

An isothermal mechanical fatigue tester that can generate a pure shear loading on the solder joints and can be tested either in displacement control mode or in force control mode is shown in Fig. 11.1. The samples are attached and tightened on holders by proper fixtures. One holder is fixed on the main body of the tester. Another holder, the movable holder, is constructed of a pair of rods, which convey force provided by a power system to the movable holder and sample (Fig. 11.2). The power system consists of a DC servomotor with tachogenerator, a belt drive, a screw with a nut and a lever-cross head. The force can be controlled and measured by power gauging that is connected to movable holder. A laser interferometer can send a red laser beam to the mirror inside the movable holder to measure the change of distance between the holders.

11.7 Shear Testing

The shear strength of solder bump or wire bond is measured by a dedicated tester, which can also be used to evaluate the effect of certain loading or atmosphere on shear strength, such as thermal aging and humidity exposure. The principle of this method is moving a shear arm to push the solder ball or wire bond off its bond pad.

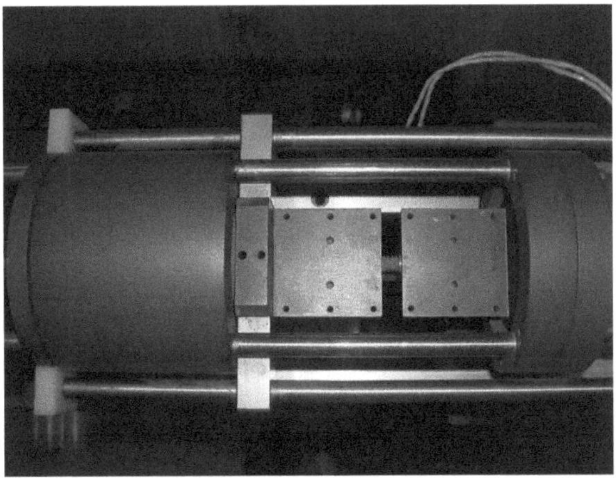

Fig. 11.2 Holders and rods

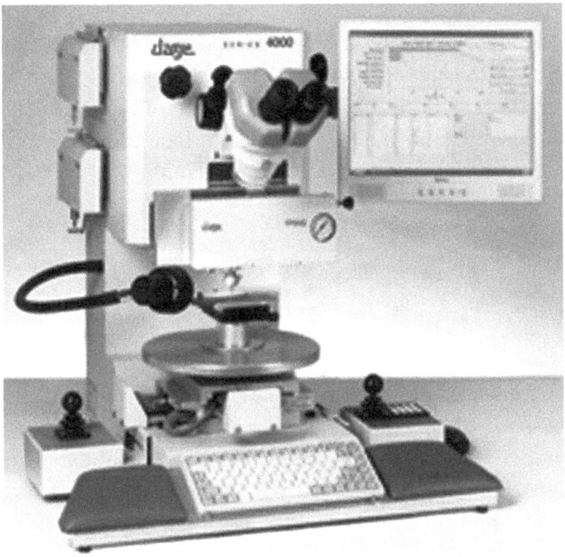

Fig. 11.3 DAGE 4000

Figure 11.3 shows an example. The basis of solder bump shear test is the removal of a solder bump from a packaging using a chisel-like shear tool with a face width comparable in size to the diameter of the test ball (Fig. 11.4). A wide variety of speeds can be used according to the application. Bond strength and failure mode can both be used to evaluate the test [1].

For conventional shear test, because of low speed limitations, the dominant failure mode tends to be solder itself, and the brittle fracture at the interface is quite scarce. This means the traditional shear test is difficult to compare the effects of different pad surface finishes, solder alloys, microstructures, and IMCs near the

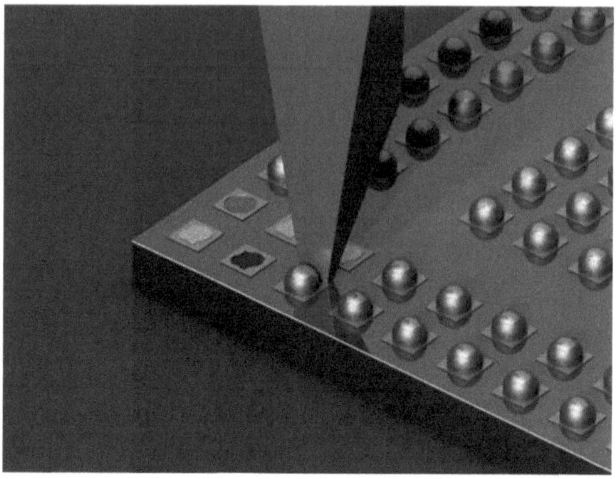

Fig. 11.4 Shear test with DAGE 4000

interface. Oppositely, high-speed shear test can produce many more interface failures, allowing different interface materials to be compared in performance. So, application of high-speed shear test (more than 1,000 mm/s) is used more and more to detect brittle fracture at the interface as an essential complement to traditional shear test (shear speed less than 20 mm/s).

11.8 Humidity and Temperature Testing

Humidity and temperature testing are examples of the accelerated aging tests. The purpose of the testing is to evaluate the reliability of electronic productions or properties of materials under high humidity and temperature. Organic materials, especially epoxy, are used widely in electronic packaging, such as, epoxy molding compound (EMC), and underfill of flip chip and BGA. Moisture has great influence on the reliability of these materials. The devices using these materials will be destroyed by cracking or warpage during the reflow process after moisture absorption. For evaluation of the reliability of electronic devices, humidity and temperature testing are used. Figure 11.5 shows the moisture absorption curves of EMC by experiment and ANSYS simulation under 85%RH/85°C and 30%RH/85°C.

The products or samples are put in the humidity/temperature chamber for a certain time. They are subjected to moisture absorption and diffusion during the process, which influences the samples' physical, mechanical, and chemical properties, and so on.

Most chambers used for the humidity and temperature testing can support the experiment for temperature from −20 to 200°C and humidity from 10%RH to 98%RH.

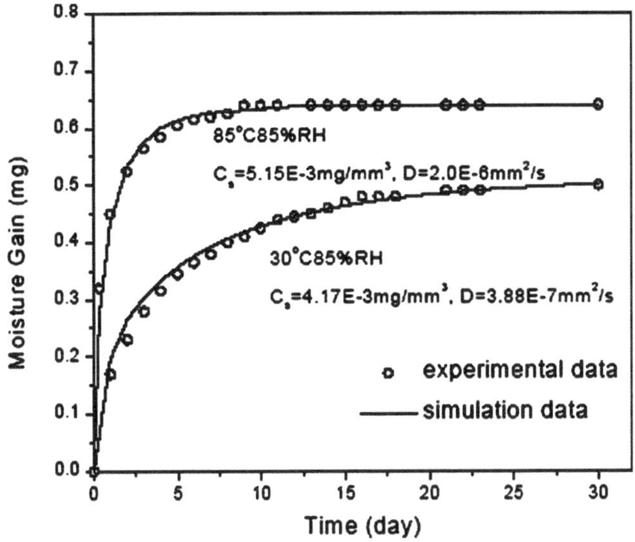

Fig. 11.5 Moisture absorption curves of experiment and simulation using ANSYS under 85%RH/85°C and 30%RH/85°C of EMC

11.9 Thermal Shock and Thermal Cycling Testing

Thermal shock and thermal cycling tests are also accelerated aging tests. During the tests, the samples are heated and cooled rapidly in a very short time. The low temperature is always below −40°C, and the high temperature is above 100°C.

Some electronic productions must work at very quick alternation between high and low temperatures. Therefore, good performance under the fast switching of very high and low temperature is desired for the devices. Thermal shock testing is used to evaluate if the products can work normally under conditions mentioned above. There are two kinds of thermal shock tests. Alternately dipping the product in hot and cold liquids is referred to "liquid-to-liquid thermal shock." Moving the product from a hot to a cold chamber or other sudden change of the air temperature is "air-to-air thermal shock" or "two-zone thermal shock." The rate of temperature change for thermal shock testing is more than 20°C/min and irregular. Failure appearing during this testing is mainly due to creep and fatigue damage.

Failure, such as crack or delamination, appears when electronic products are subject to work under alternation between high and low temperatures. To evaluate the reliability of electronic products, thermal cycling is used. Thermal cycling changes the air temperature in a single chamber. The rate of temperature change is commonly less than 20°C/min to avoid the influence of thermal shock, and it is controllable and regular during the testing. The failures appearing during this testing result mainly from shear fatigue. Figure 11.6 is a curve of thermal cycling used for electronic devices.

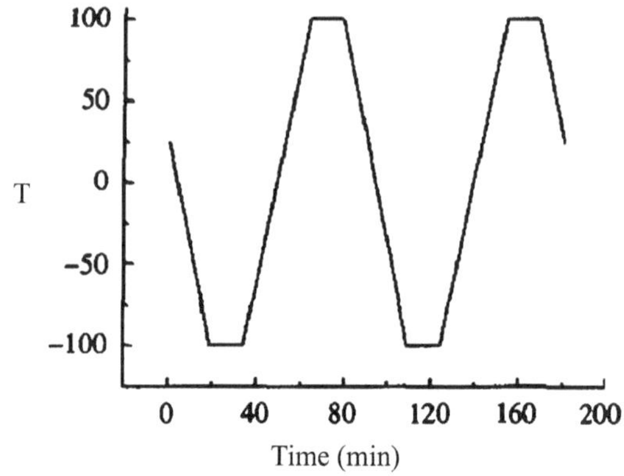

Fig. 11.6 Temperature vs. time curve of thermal cycling testing

The main part of the chambers used for thermal shock and thermal cycling testing is a big compressor for reaching the very low temperature rapidly.

11.10 Moiré Interferometry

As microelectronics devices are made smaller, the thermal gradient increases, and the strain concentrations become more serious [1]. While numerical analyses have been used extensively to estimate stresses and strains in packaging structures, advanced experimental techniques are in high demand to provide accurate solutions for deformation studies for microelectronics devices.

Among various experimental methods, moiré interferometry is an optical measurement method. The optical arrangement and operation mechanism of the interferometer can be found in detail in [2]. Moiré interferometry can provide whole-field contour maps of displacements with subwavelength sensitivity and give abundant displacement data, which permit reliable determination of normal and shear strains.

Moreover, the moiré interferometer measurement is characterized by a list of excellent qualities, including the following:

- Real-time technique: the displacement fields can be viewed as loads are being applied.
- High sensitivity to displacements, and higher for microscopic analyses.
- High spatial resolution: measurements can be made in tiny zones.
- High signal-to-noise ratio: the fringe patterns have high contrast and excellent visibility.
- Large dynamic range: the method is compatible with a large range of displacements, strains, and strain gradients.

These features make the methods ideally suited for the cases with a complex geometry. Especially, moiré interferometry measurements with regard to the

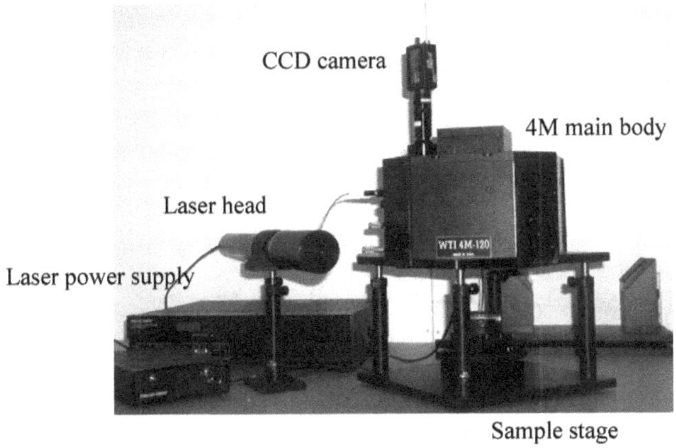

Fig. 11.7 Multifunction Macro Micro Moiré Interferometer

packaging technology show the applicability of this method. Han et al. [3, 4] made a thermomechanical deformation analysis. The deformation fields of solder balls in BGA flip-chip packaging under different thermal loadings have been observed [5–8]. Shi and Wang [9] used the moiré Interferometer to study the thermal deformation of the solder ball as well as the interfacial behaviors of a copper–solder interface. Ham et al. [10, 11] investigated the electronic conductive film deformation under thermal cycling.

The instrument shown in Fig. 11.7 is an in-plan Multifunction Macro Micro Moiré Interferometer (4M), which is a variation of the four-beam moiré interferometer. The schematic diagram of the 4M optical arrangement is illustrated in Fig. 11.8.

In the method, a high frequency crossed-line diffraction grating is replicated on the surface of the specimen and deforms together with the underlying specimen. 4M is a compact system that utilizes a cross-line grating and various mirrors to produce the four incident beams. A diverging laser beam from a fiber tip is directed by the 45° mirrors to pass through a collimation lens L2, and then the collimated beam strikes a cross-line diffraction grating G (usually with a frequency of 1,200 lines/mm) at normal incidence. Light diffraction by the specimen grating is collected by the camera lens, which focuses images of the moiré patterns onto the record plane of a camera or CCD target. In practice, two opposite beams are blocked, while the other two produce the fringe pattern.

Isothermal loading causes deformation of the assembly components, which can be obtained from the fringe patterns. The 4M interferometer is capable of measuring in-plane displacements with very high sensitivity.

During the experimental observation, the interferometer produces the moiré fringes when the deformed specimen grating interferes with the virtual reference grating. The resulted fringe patterns generate the contour maps of the displacement fields.

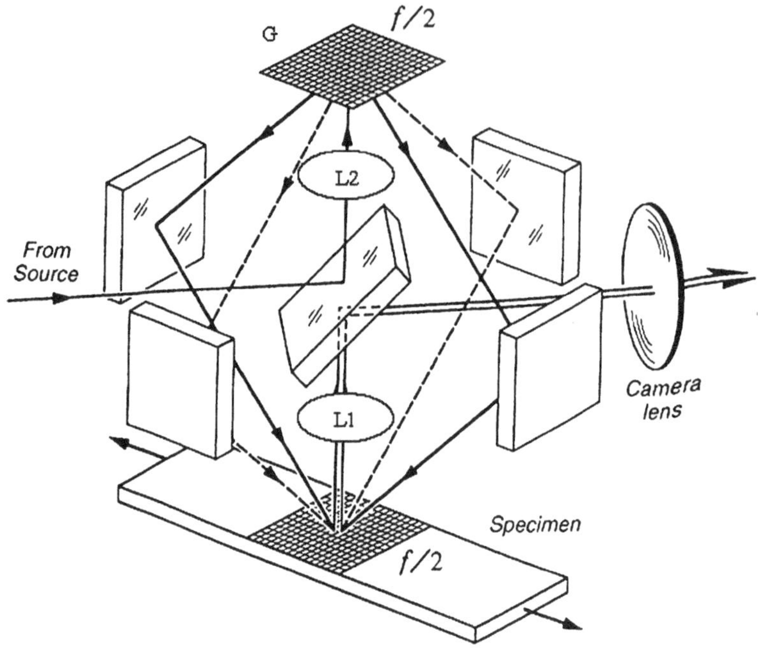

Fig. 11.8 Schematic diagram of optical arrangement of 4M

The relationship between fringe order and displacement is as follows:

$$U = \frac{1}{f} N_x, \tag{11.1}$$

$$V = \frac{1}{f} N_y, \tag{11.2}$$

where f is the frequency of the virtual reference grating, U is the displacement in the x-direction in the measurement plane, N_x is the fringe order in the x-direction, V is the displacement in the y-direction in the measurement plane, N_y is the fringe order in the y-direction.

When strains are required, they can be extracted from the displacement fields by the relationships for engineering strains:

$$\varepsilon_x = \frac{\partial U}{\partial x} = \frac{1}{f}\left[\frac{\partial N_x}{\partial x}\right], \tag{11.3}$$

$$\varepsilon_y = \frac{\partial V}{\partial y} = \frac{1}{f}\left[\frac{\partial N_y}{\partial y}\right], \tag{11.4}$$

and

$$\gamma_{xy} = \frac{\partial U}{\partial y} + \frac{\partial V}{\partial x} = \frac{1}{f}\left[\frac{\partial N_x}{\partial y} + \frac{\partial N_y}{\partial x}\right], \tag{11.5}$$

Besides the in-plane deformation measurement, there is shadow moiré, which can obtain the out-of-plane displacement of the studied specimen. The out-of-plane interferometer is mainly used for measuring the warpage of electronic products under thermal processes.

As stated earlier, superposing two periodic images generates moiré patterns. In the shadow moiré interferometer, one image comes from a glass grating used as the reference, and the other is the shadow of the grating lines on a surface being measured. Small variations from the reference grating are magnified by moiré fringes and give a quantitative measure of surface topology. A CCD camera captures the moiré patterns. Then the out-of-plane displacement with respect to the reference grating can be interpreted [12].

Exercises

11.1 Consider and try to explain why does shear speed have an obvious effect on failure mode?

11.2 What is the difference between thermal shock testing and thermal cycling testing?

11.3 Calculate the displacement of the solder joint in the following picture (here the virtual frequency is $f = 2{,}400$ line/mm).

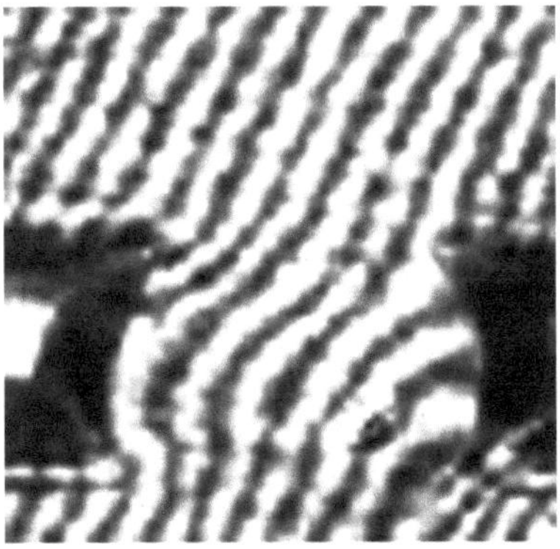

11.4 How can we find the glass transition temperature of a conductive adhesive system? Explain the mechanism.

11.5 In which way(s) can we do accelerated aging tests?

References

1. R. Tummala, Fundamentals of Microsystem Packaging, New York: McGraw-Hill, 2001.
2. D. Post, J. D. Wood, "Determination of Thermal Strains by Moiré Interferometry", Experimental Mechanics, 29(3), 1989, 318–322.
3. B. Han, "Recent Advancements of Moiré and Microscopic Moiré Interferometry for thermal Deformation Analyses of Microelectronics Devices", Experimental Mechanics, 38(4), 1998, 278–288.
4. B. Han, D. Post, "Immersion Interferometer for Microscopic Moiré Interferometry", Experimental Mechanics, 32(1), 1992, 38–41.
5. J. W. Joo, K. W. Oh, S. M. Cho, B. Han, "Thermo-Mechanical and Flexural Behavior of WB-PBGA Package Using Moiré Interferometry", Proceedings of 3rd International Symposium on Electronic Materials and Packaging, Cheju Island, Korea, 2001, 421–427.
6. J. S. Zhu, D. Q. Zou, F. L. Dai, S. Liu, "High Temperature Deformation of High Density Interconnects and Packages by Moiré Interferometry/FEM Hybrid Method", Proceedings of 19th IEEE/CPMT International Electronics Manufacturing Technology Symposium, Austin, TX, USA, 1996, 75–83.
7. S. Liu, J. J. Wang, D. Q. Zou, X. Y. He, Z. F. Qian, "Resolving Displacement Field of Solder Ball in Flip-Chip Package by Both Phase Shifting Moiré Interferometry and FEM modelling", Proceedings of 48th Electronic Components and Technology Conference (ECTC), Seattle, WA, USA, 1998, 1345–1353.
8. E. A. Stout, N. R. Sottos, A. F. Skipor, "Mechanical Characterization of Plastic Ball Grid Array Package Flexure Using Moiré Interferometry", IEEE Transactions on Advanced Packaging, 32(4), 2000, 637–645.
9. X. Q. Shi, Z. P. Wang, "In-Situ Moiré Interferometry Technique and Its Applications to Microelectronic Packages", Proceedings of 52nd Electronic Components and Technology Conference, San Diego, USA, 2002, 183–191.
10. W. S. Kwon, S. J. Ham, K. W. Paik, "Deformation Mechanism and Its Effect on Electrical Conductivity of ACF Flip Chip Package under Thermal Cycling Condition: An Experiment Study", Microelectronics Reliability, 46(2–4), 2006, 589–599.
11. T. K. Hwang, S. J. Ham, S. B. Lee, "A Study on the Thermal Deformation of ACF Assemblies Using Moiré Interferometry and FEM", International Symposium on Electronic Materials & Packaging, Hong Kong, China, 2000, 358–363.
12. H. Ding, R. E. Powell, C. R. Hanna, I. Charles Ume, "Warpage Measurement Comparison Using Shadow Moiré and Projection Moiré Methods", IEEE Transactions on Components and Packaging Technologies, 25(4), 2002, 714–721.

Abbreviations

ACA	Anisotropic conductive adhesive
ALT	Accelerated life testing
ASIC	Application-specific integrated circuits
BCT	Body-centered tetragonal
BGA	Ball grid array
CTE	Coefficient of thermal expansion
DSC	Differential scanning calorimetry
ECA	Electrically conductive adhesive
EDX	Energy dispersive X-ray
ESD	Electrostatic discharge
FEA	Finite element analysis
FIT	Failures-in-time
FR-4	Flame retardant-4
HASS	Highly accelerated stress screens
ICA	Isotropic conductive adhesive
LCF	Low cycle fatigue
MEMS	Microelectromechanical systems
MTTF	Meantime-to-failure
NCF	Nonconductive film
OM	Optical microscopy
PCB	Printed circuit board
PDF	Probability density function
PSB	Persistent slip band
PWB	Printed wiring board
RBD	Reliability block diagram
RH	Relative humidity

J. Liu et al., *Reliability of Microtechnology: Interconnects, Devices and Systems*,
DOI 10.1007/978-1-4419-5760-3,

SAM	Scanning acoustic microscopy
SAC	Sn–Ag–Cu
SEM	Scanning electron microscopy
SIR	Surface insulation resistance
SMT	Surface mount technology
SOP	System on package
TEM	Transmission electron microscopy
UBM	Under bump metal

Answers to the Exercises

Chapter 2

2.1. Graph shows failure occurrences of a product population within a time. The graph can be divided into three distinct areas; first, the products with inadequate quality will fail. Then, the product failures are initiated from random sources, e.g., from out-of-specification usage. In the third area, the failure occurrence is increasing, which is caused by wear out mechanisms of components or interconnections.

2.2. Reliability of each interconnection = 81%.

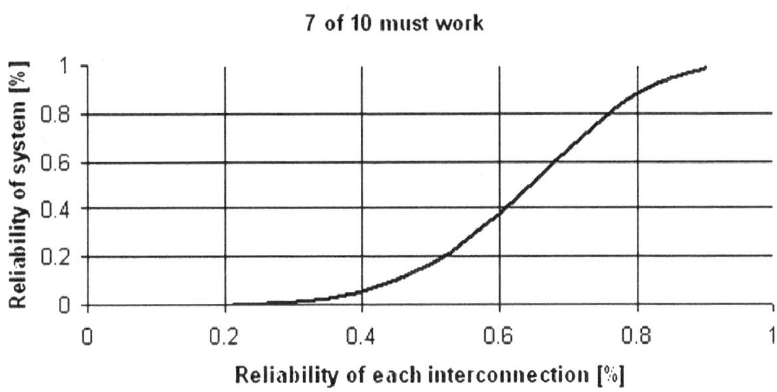

2.3. Draw the bathtub curve. It can be seen that the failure is high in the beginning due to unoptimized design. Then it comes into the fatigue stage with constant failure rate. Finally, we enter into the aging stage with accelerated failure.

2.4. First determine the cubic functions of the bathtub curve:

Infant mortality:	$y(t) = 2 \times 10^{-10}[(200 - t)^3 + 2 \times 10^6]$	$0 \le t \le 200$
Bottom	: $y(t) = 4 \times 10^{-4}$	$200 \le t \le 800$
Wearout	: $y(t) = 2 \times 10^{-10}[(t - 800)^3 + 2 \times 10^6]$	$800 \le t$

Then integrate to find

$F(t) = 2 \times 10^{-10}[\{200^4 - (200 - t)^4\}/4 + 2 \times 10^6\, t] \qquad 0 \leq t \leq 200$

$F(t) = 4 \times 10^{-4}(t - 200) + 0.16 \qquad 200 \leq t \leq 800$

$F(t) = 2 \times 10^{-10}[(t - 800)^4/4 + 2 \times 10^6(t - 800)] + 0.4 \quad 800 \leq t$

a) Solve the last equation for t when $F(t) = 1$, gives approx: 1110 hours.
b) 1110 hours. Read from Fig. 8 when $F(t) = 1$.

2.5. Using MATLAB for calculation, plotting, and parameter extraction:

WEIBULL:

Table A		Table B	
eta	= 5.868337e+02	eta	= 3.571399e+02
beta	= 1.795954e+00	beta	= 1.023981e+00
mttf	= 5.219207e+02	mttf	= 3.536840e+02

LOG-NORMAL

Table A		Table B	
mu	= 6.019587e+00	mu	= 5.336802e+00
sigma	= 8.440636e−01	sigma	= 1.161794e+00
mttf	= 5.874602e+02	mttf	= 4.081707e+02

See plots below:
The Weibull plot looks better for both data sets, but the fit is not perfect.

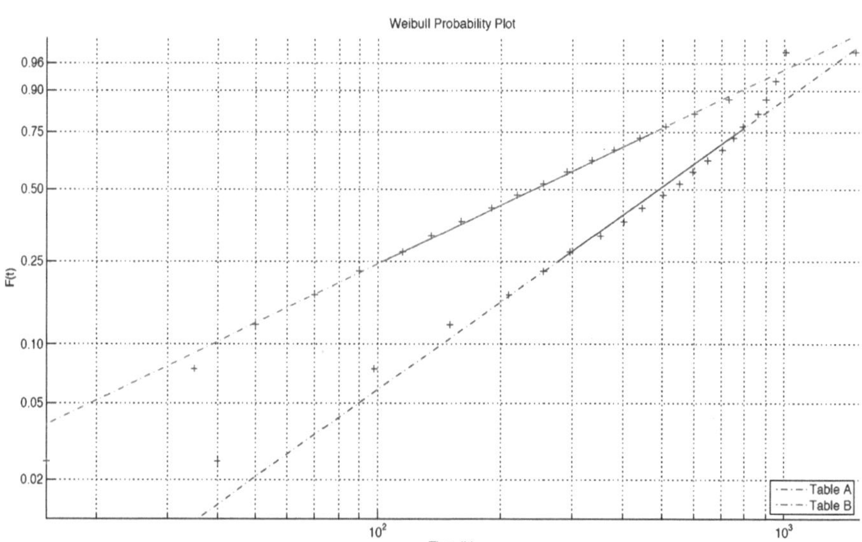

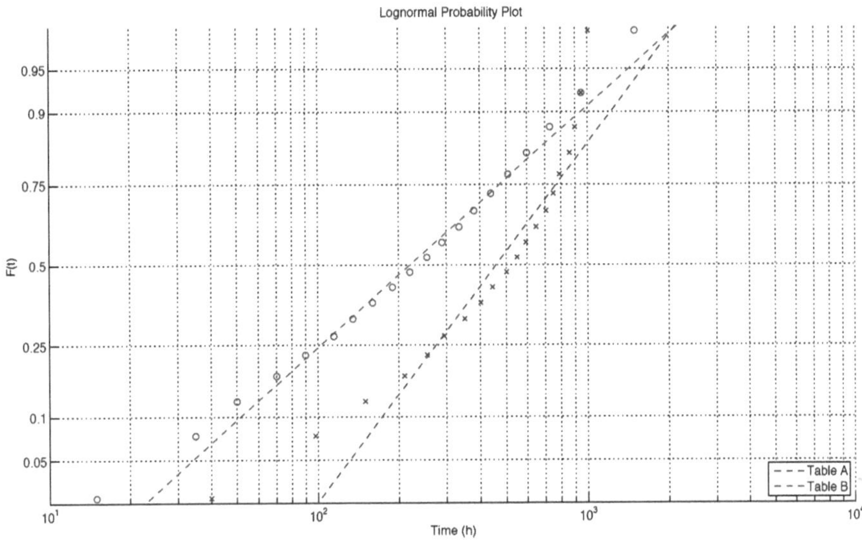

Chapter 3

3.1.

Category	Failure cause	Description/examples
Hardware	Parts	ICs, transistors, resistors, connectors, etc.
	Manufacturing	Anomalies in the manufacturing process, i.e., solder joint defects, etc.
	Wear out	Component-related examples are drying electrolytic capacitors and switch wear out
	Induced	External applied stress, i.e., dropping, bending, electricity, etc.
Software	Software	Failures of a system to perform its intended function due to the manifestation of software fault
Design	Design	Failures resulting from an inadequate design, i.e., tolerance stack-up, unanticipated logic conditions
	System management	Failures related to faulty interpretation of system requirements or errors in the subpart interfaces, etc.
No failure	No defect	Perceived failures that cannot be reproduced upon further testing

Also see Fig. 3.2 in the text book.

3.2. The definition of failure can be phrased as "any condition that causes a device or circuit to fail to operate in a proper manner."
Failure causes of an electronic system can generally be divided into three categories: hardware, software, and design-related failures.

3.3. There are many reasons for this of which many are human error related. For instance, the product was not used within its specification and did not function properly due to that. Also, if the failure cannot locate so that the product could be repaired, it could be labeled as "no failures found."

3.4. Temperature changes are initiated from (1) switching on/off, (2) power consumption based on usage, and (3) atmospheric changes among others. Thermal stress is a primary cause for a fatigue crack in solder joints.

3.5. Brittle fractures are typically addressed to incompatible materials in excess stress conditions, i.e., in drop conditions. Fracture occurs in the brittle intermetallic layers of solder joint. The brittle fracture does not evolve as a result of plastic deformation, but is an instant rupture through material or several materials due to excess external stress. Breaking glass is a good example of such a failure event, in which the failure could easily be detected, but no evidence when it was going to happen could have been recorded. In electronic brittle factures occur typically in shock loading environments, for example as a result of drop, where over than 1,000 Gs can be generated. Brittle fracture is typically due to incompatible material selections.

3.6. The fatigue failure mechanism can be divided into three phases: (1) crack nucleation, (2) crack propagation, and (3) final fracture. The crack nucleation is preceded by microstructural changes, e.g., local grain growth. The crack propagation starts with crystallographic propagation, which is followed by noncrystallographic propagation. After enough plastic deformation has taken place, the mechanical strength and electrical path integrity of the solder joint has totally gone, meaning that final fracture has been fully developed. The solder fatigue failures typically occur suddenly and unexpectedly as no observable plastic deformation occur before the failure.

3.7. In principle, the metal deformation under static load can be divided into multiple phases, called primary creep, secondary creep, and tertiary creep. At the primary creep strain stage, any microstructural evidence of creep damage can be found from the material. The secondary creep is also known as steady-state creep, as the strain level will maintain relatively constant. At this stage, the work hardening rate is balanced by thermally activated recovery rate. Individual voids start to occur at the microstructure level. At the tertiary creep region, the material experience higher strain rates than at the secondary creep. At this stage, the voids start to grow and to form cracks that will end up to final rupture.

3.8. By proper IC design. Factors that play a major role in electromigration are (1) temperature and its gradient, (2) current density, (3) conductor dimensions, (4) conductor grain structure, and (5) impurities.

3.9. When the surfaces of two materials closely glide against each other, they are electrically charged. Depending on the molecular structure of the materials, there is a tendency for one surface to strip electrons from the other. This is commonly called as triboelectric charging. An everyday example of the triboelectric charging is when a person walks across the carpet and is then electrically charged up to 20 kV. When person touches an object that is at different electrical charging level, the charge will be leveled between the person and the object. The discharging takes place within less than 100 ns. If the discharge path includes weak parts, e.g., material layers in nanometer scale, etc., the relatively high current density will result as dramatic deterioration in the material microstructure.

3.10.

Mechanism	Description
Bulk breakdown	Transistor parameter shifting due to breakdown in transistor microstructure. Breakdown path goes from Al-electrode through doped regions (P- or N-type) to silicon substrate. High current density causes alloying of semiconductor and precipitation of Al in the doped regions
Thermal secondary breakdown	Leakage current increase due to breakdown between PN-junction due to high voltage. This high speed current pulse generates very high temperature increase locally. Due to this, it damages the structure only in limited volume
Surface breakdown	Short circuiting or increase in leakage current due to breakdown between two adjacent metal conductors. The breakdown path progresses typically on the dielectric surface, hence the name
Dielectric breakdown	Short circuiting or increase in leakage current as a result of a high voltage breakdown through the dielectric material
Electromigration	Opens caused by a high electrical current density, which moves the atoms of the conductor in the direction of the electrons
Latch up	Transistor latches up due to ESD pulse in transistor microstructure. This is due to undesirable biasing of PN-junctions

3.11.

	Preventive action
1	Use only electrically grounded working places with static dissipative materials
2	Avoid or minimize the using or existence of charging materials at production floor
3	Make sure that the whole handling chain of the components or other subparts of the product, including human and machinery operations and movements, is according to ESD protection policy
4	Operators must always use grounding wrists or heel straps when handling ESD sensitive products
5	Minimize handling
6	Give feedback to product design responsibilities to avoid the usage of ESD sensitive components or to take the ESD into account in the product design

3.12. Moisture (H_2O) is playing a major role in corrosion mechanisms. Moisture is present in air and is then constantly present in electronics.
The galvanic corrosion is of particular concern in solder joints.
At IC microstructure level, there are several corrosion-induced failure mechanisms. Examples of these are (1) corrosion of aluminum metallization or wire, (2) corrosion of intermetallic compounds, (3) corrosion of gold wires, (4) corrosion of copper wire, and (5) corrosion from die bond material.

3.13. With proper product design and material selections.

3.14. It is about your story.

Chapter 4

4.1. Lead has been identified as one of the most toxic elements in the world. It accumulates in the body and central nerve systems and causes damages in terms of decreased learning capacity for instance.

4.2. Lead (Pb) in semiconductor products has come under increased environmental scrutiny because of the growing number of electronic products requiring end-of-life treatment and disposal. A variety of jurisdictions around the globe have proposed regulations that would restrict the use of Pb or impose additional requirements when Pb is used in products.

4.3. There is a law in EU and many other countries to ban the lead in electronics in general with some exceptions.

4.4. Environmentally compatible. High creep strength in general compared with Pb containing solder. The European Community has established a phase out date of July 1, 2006 for Pb in electronic products, with some exceptions.

4.5. Solder that contains less than 0.2% lead. Lead-free solder has different properties and appearance than lead-based solder. Lead-free solder is dull and grainy and requires hotter soldering temperatures.

4.6. Higher melting of new lead-free solders leads to higher reflow temperature, which may potentially cause component damage. Uncertain about the long-term reliability due to limited field failure data. New mechanisms of failure need to be discovered.

4.7. We need to use Pb-free finishes on substrate. Higher reflow temperature such as 250–260°C is required.

4.8. No, not necessary. Basically, the printing parameters and stencil design are determined by the rheology and particle size as well as pitch requirement.

4.9. Higher soldering temperature leads to higher vapor that builds up inside the cavity with moisture. Therefore, higher temperature may lead to the formation of pop-corning effect easier.

4.10. Thermomechanical design means designing a product taking into account the problems that can occur by thermomechanical mechanisms. It is important to understand why a specific failure mechanism happens and to design the product against such mechanism.

Generally, there is two ways to design against failures, either by reducing the stresses that can cause the failure or by increasing the strength of the "component." Either, or, the important thing is to avoid failure.

But what causes thermomechanical failures?

This failure mechanism is caused by stresses and strains generated within an electronic package due to thermal loading from the environment or internal heating in service operation. Due to the CTE mismatch among different materials, and due to thermal gradients in the system, and due to geometric constrains, thermally induced stresses and strains are generated in various parts of a system.

Example of concrete measurements:

- Use similar CTE
- Shorter DNP (distance to neutral point)

- Materials with high T_m (⇒If the homologous temperature $T_h = T/T_{melt}$ is over 0.5, the risk for creep is significant).

4.11. The CTE dependency on temperature says that the higher the temperature the higher the CTE.
The higher the CTE a material has higher deformations, this material is going to exhibit, which means that if the CTE mismatch between the different parts in a package is larger, the higher is the risk for failure. In reality, we desire materials with low CTE that are more stable and do not change dimensions (not so much) with changes in temperature. In addition, if we have materials with similar CTEs, then the CTE mismatch would be small and the solders would not become too stressed. That means the package will have higher thermomechanical reliability.
The higher the yield stress Sy(T), the higher the materials elasticity limit. That means that the material withstands higher stresses and strains without any permanent deformation.
Bellow the yield point, the material is in the linear elastic region and the elastic deformation vanishes when the applied load is removed. Above the yield point, the stress–strain relationship can be described as a nonlinear function, and in this region, the plastic deformation is permanent and does not vanish when the applied load is removed.
For reliability, and to avoid plasticity, one should choose materials with high yield stress Sy(T). Even in this case, the higher the yield stress of the materials in a package, the higher the thermomechanical reliability.
The elastic modulus dependency on temperature states that the higher the temperature the lower the elastic modulus.
The elastic modulus E can be thought as a material's stiffness or a material's resistance to elastic deformation. The greater the modulus the stiffer the material or the smaller the elastic strain that results from the application of a given stress. The elastic deformation is nonpermanent, which means that when the applied load is released, the piece returns to its original shape. On an atomic scale, macroscopic elastic strain is manifested as small changes in the interatomic spacing and the stretching of interatomic bonds. From an atomic point of view, plastic deformation corresponds to the breaking of bonds with original atom neighbors and then reforming new bonds with new neighbors.

Chapter 5

5.1. From the curve, we can see that the shrinkage is about 3 μm, then the shrinkage is 3/60 = 5%.

5.2. Offer higher resolution capability than present solder pastes due to smaller particle size. Cured at much lower temperatures than used for soldering (thermally sensitive components and substrates). Nonsolderable (cheap) substrates can be used (e.g., glass, polyester flex). Less process steps are needed than (wave) soldering (no temporary SMD adhesive, no flux, no flux cleaning).

5.3. $Z = 3.23 \times 10^{14}\,\text{s}^{-1}, \quad E_a = 106.4\,\text{kJ}, \quad R = 8.314\,\text{J/mol/K},$

$n = 2, \quad T = 140 + 273 = 413\,\text{K}, \quad x = 99.99\%,$

$$\frac{dx}{dt} = Z \exp\left(-\frac{E_a}{RT}\right)(1-x)^2,$$

$$\frac{1}{(1-x)^2}dx = Z \exp\left(-\frac{E_a}{RT}\right)dt,$$

$$\int \frac{1}{(1-x)^2}dx = \int Z \exp\left(-\frac{E_a}{RT}\right)dt + C,$$

$$(1-x)^{-1} = Z \exp\left(-\frac{E_a}{RT}\right)t + C,$$

when $(t = 0), \quad x = 0,$

so, $C = 1,$

$$(1-x)^{-1} = Z \exp\left(-\frac{E_a}{RT}\right)t + 1,$$

$$(1-0.9999)^{-1} = 3.23 \times 10^{14} \exp\left(-\frac{106.4 \times 10^3}{8.314 \times 413}\right)t + 1,$$

$t = 865.71\,\text{s}.$

5.4. Morrow's law: $N_f^m W_p = C$, where m is the fatigue exponent and C is the material ductility coefficient, W_p is the strain energy density. It takes into account both stress and strain.
Coffin–Manson relationship: $N(\Delta\varepsilon_p)^n = C_{pf}$, where N is the number of cycles to failure, n is an empirical constant, $\Delta\varepsilon_p$ is the plastic strain range during one cycle, and C_{pf} is a proportionality factor. It takes into account only plastic strain.

5.5. For flip-chip solder joining, plastic strain of solder bumps is a critical parameter that governs the joint reliability. Using a high bump can reduce the bump strain and thus increase the joint reliability, as shown in Fig. A.a below. However, a systematic study of the effect of bump height showed that the failure mechanism of ACA flip-chip joints is totally different. In ACA joints, the bump and pad are usually made of metals that are much stiffer than adhesives. In other words, thermal mismatch stresses can hardly deform the bump and pad, and the shear strain is localized in the adhesive between the mating bump and pad as shown in Fig. A.b below. In this case, the joint reliability is governed by the shear strain in the adhesive and the influence of bump height is limited. Meanwhile, the stress in the Z-axis will be raised with bump height due to the increased adhesive volume. At elevated temperature, this stress can lift the chip and weaken the joint. So benefits from high bumps cannot be expected for ACA joints.

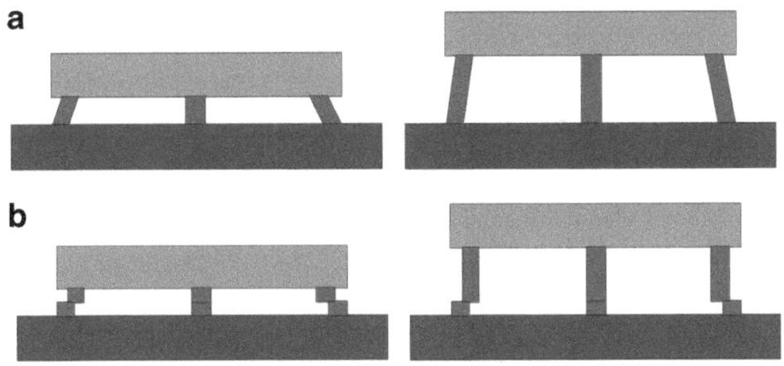

Chapter 6

6.1. **Case 1:** without underfill

In this case, we use the model depicted below.

$$\Delta\gamma = \Delta U/h = (d \cdot \Delta T \cdot \Delta\alpha)/h = (0.003 \times 180 \times 15.7)/0.001 = 0.085$$

$$N_f = 0.5 \times (\Delta\gamma/2\varepsilon f)^{1/c}$$

One thing that we do not know is the cycle time. I assume a cycle time of 40 min $\Rightarrow f = (24 \times 60)/40 = 36$ cycles/day

$c = -0.442 - (6 \times 10^{-4}) \times T_m + (1.74 \times 10^{-2}) \times \ln(1 + f)$

$c = -0.442 - (6 \times 10^{-4}) \times 35 + (1.74 \times 10^{-2}) \times \ln(1 + 36)$

$c = -0.400170028$

$$N_f = 0.5\left(\frac{\Delta\gamma_T}{2\varepsilon'_f}\right)^{1/c},$$

$N_f = 0.5 \times (0.085/0.65)^{-2.499} = 80.7$ cycles.

Case 2: with underfill

In this case, one can use FEM to calculate the strain. Moiré analysis could also be used.

It is easy to underestimate the role of thermal radiation as a significant contributor to electronics cooling in environments without forced air flow. By its very nature it is invisible. The proper treatment of it can be intimidating due to the complicated nature of the phenomenon in environments in which localized hot regions are in the view of other localized hot regions.

However, it is possible to get a basic understanding of radiation without even worrying about such complications as view factors. The first thing to do is to respond to the basic engineering urge to linearize anything possible. Hence, (A.1) is a recasting of the familiar Stephan–Boltzmann equation, dividing it by the temperature difference between a surface (assumed isothermal) and the facing surface (assumed to be at the air temperature). The result is a heat transfer coefficient, which represents the effect of radiation at a given temperature.

$$h_{\text{RAD}} = 5.67 \times 10^{-8} \varepsilon \left(T^2_{\text{surface}}(K^2) + T^2_{\text{air}}(K^2)\right) \cdot (T_{\text{surface}}(K) + T_{\text{air}}(K))\ \text{W/m}^2\,\text{K}. \tag{A.1}$$

The numerical factor is the Stephan–Boltzmann constant and ε is the emissivity. The emissivity is in the range 0.8–0.9 for dielectrics and 0.1–0.2 for commercial metals. The temperatures are expressed in absolute temperature Kelvin units.

Even though we have linearized the S–B equation, the resultant heat transfer coefficient is still highly temperature dependent. In fact, it is proportional to the third power of the absolute temperature. Figure A.1 illustrates this temperature dependence, where we have assumed an emissivity of 0.8 and a temperature difference between the surface and the air of 1°C.

The lower x-axis indicates absolute temperature. The upper x-axis indicates degrees centigrade in the range of interest to electronics cooling. At a typical ambient temperature range, say around 50°C, h_{RAD} is approximately 6 W/m^2 K.

It is useful to compare the radiation heat transfer coefficient to the heat transfer coefficient applicable to a horizontal printed circuit board in a large enclosure. This expression represents an average for heat transfer from the top and bottom surfaces of the board.

$$h_{\text{NC}} = 3.76(\Delta T_{\text{surface-air}}(K))^{0.25}\ \text{W/m}^2\,\text{K}. \tag{A.2}$$

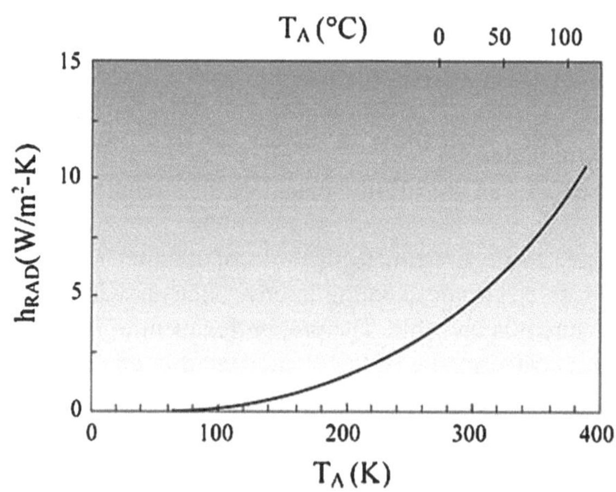

Fig. A.1 Temperature dependence

The following graph, Fig. A.2, compares the magnitude of the radiation and natural convection heat transfer coefficients as a function of the temperature difference between the surface and air temperature, where the air temperature is assumed to be 50°C.

One sees that the h_{RAD} is actually greater than h_{NC} up to a temperature difference of about 25°C. For temperature differences exceeding this, they are nearly equal.

In more realistic situations, the details of radiation heat transfer can be very complicated. The relative heat transfer by radiation and natural convection can differ significantly from that demonstrated in this comparatively simple example. However, the fact remains that radiation heat transfer is significant in many natural convection cooling situations and must not be overlooked.

6.2.

1. The homologous temperature for eutectic solder:

$$T_h = \frac{(273 + 23)}{(273 + 180)} = 0.653421633 \approx 0.65.$$

2. The Darveaux's constitutive law for eutectic solder:

$$f(\tau)g(t)h(T) = C\left[\sinh\left(\varpi\frac{\tau}{G}\right)\right]^n \times t \times \left(\frac{G}{T}\right)\exp\left(\frac{-Q}{kT}\right).$$

3. The steady-state creep shear strain rate

$$\frac{dx}{dt} = \frac{(f(\tau)g(t)h(T))}{dt} = C\frac{G}{T}\left[\sinh\left(\varpi\frac{\tau}{G}\right)\right]^n \exp\left(\frac{-Q}{kT}\right).$$

6.3.

Question 1: What is the shear strain imposed on the gold bumps when it has no underfill?

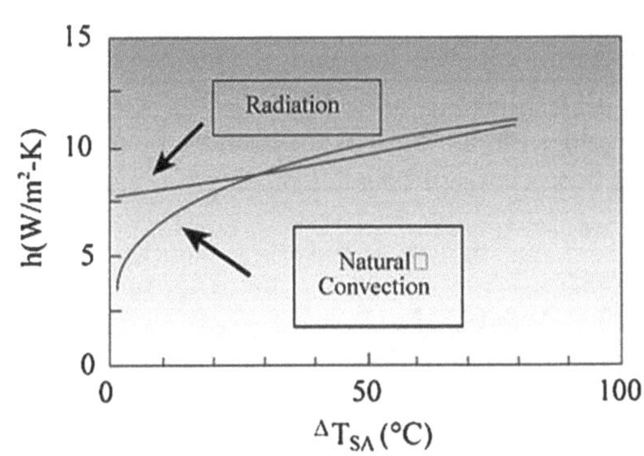

Fig. A.2 Magnitude of radiation and natural convection heat transfer coefficients

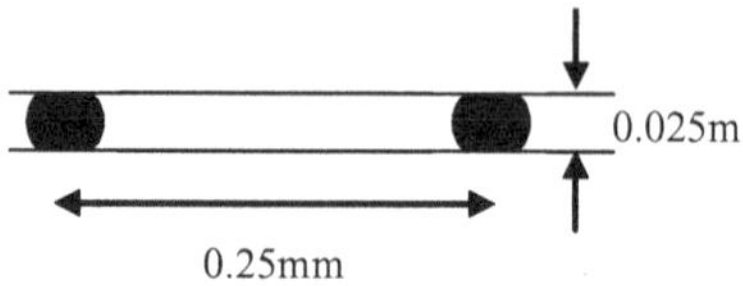

Gold bump

DNP = d = (25 × 0.25)/2 = 3.125 mm
$\alpha_{ceramic} - \alpha_{chip} = 10-5 = 5 \times 10^{-6}$/K
$\Delta T = 125 + 40 = 165$ K
$h = 0.025$ mm

$\gamma = \Delta U/h = (\mathrm{d}\cdot\Delta T\cdot\Delta\alpha)/h = (3.125 \times 165 \times 5 \times 10^{-6})/0.025 = 0.103125.$

Question 2: Consider the eutectic solder bump, which is farthest away from the chip center (assuming the distance equals half of the ceramic substrate edge), what is the shear strain on it?

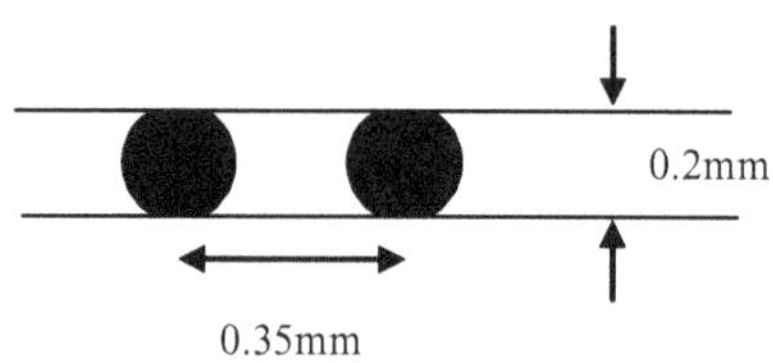

63Sn/37Pb Solder Ball

DNP = d = 3.725 mm (assuming the distance equals half of the ceramic substrate edge)
$\alpha_{PWB} - \alpha_{ceramic} = 16 - 10 = 6 \times 10^{-6}$/K
$\Delta T = 125 + 40 = 165$ K
$h = 0.2$ mm
$\gamma = \Delta U/h = (d\cdot\Delta T\cdot\Delta\alpha)/h$
$= (3.725\ [\mathrm{mm}] \times 165\ [\mathrm{K}] \times 6 \times 10^{-6}\ [/\mathrm{K}])/0.2\ [\mathrm{mm}] = 0.01843875.$

Question 3: Assuming both gold bumps and eutectic bumps in this study obey the same Coffin–Manson equation with the exponential number −2, the lifetime for solder bumps is what times more than that of gold bumps? (This situation has been changed because of the existence of underfill.)

$N_f = C(\Delta\gamma)^{\beta}$
With $\beta = -2$
Gold bumps: $\Delta\gamma = 0.103125$
Eutectic bumps: $\Delta\gamma = 0.01843875$
$N_f = C(0.103125)^{-2} = C \times 94.0312213$
$N_f = C(0.01843875)^{-2} = C \times 2941.284593$

The life of the solder bumps is going to be 31 times larger than the life of the gold bumps.

Question 4: An interesting phenomenon is observed in experiments. For the eutectic bumps on PCB, the farer distance from the chip center, the easier it will be damaged. Can you explain it?

$\Delta U = d \cdot \Delta T \cdot \Delta\alpha$, where $d = \text{DNP}$.

Also, the greater the DNP the greater the displacement imposed on the solder bump. That results in a higher force applied to the solder bump, which results in damage.

6.4.

$\Delta U = \mathrm{d} \cdot \Delta T \cdot \Delta\alpha$, where ΔU is the relative displacement, d is the distance to neutral point (DNP), ΔT is temperature difference, and $\Delta\alpha$ is thermal coefficient difference.

Shear strain: $\gamma = \Delta U/h = (d \cdot \Delta T \cdot \Delta\alpha)/h$, where h is the stand-off height. In this case, we have $\Delta T = 100°\text{C}$, $\Delta\alpha = 20 - 7 = 13$ ppm/°C $= 13.10 - 6$/°C, $d = \text{DNP} = 11.4$ mm.

$\Delta U = d \cdot \Delta T \cdot \Delta\alpha = 11.4\ [\text{mm}] \times 100\ [°\text{C}] \times 13.10^{-6} [/°\text{C}] = 0.01482$ mm

6.5. Engelmaier's model takes in to account both the plastic strain range and the frequency. It says that:

$$N_f = 0.5\left(\frac{\Delta\gamma_{\mathrm{T}}}{2\varepsilon_f{}'}\right)^{1/c},$$

where $c = -0.442 - (6 \times 10^{-4}) \times T_{\mathrm{m}} + (1.74 \times 10^{-2}) \times \ln(1 + f)$, T_{m} is mean cyclic temperature (°C), f is cyclic frequency ($1 \leq f \geq 1{,}000$ cycles/day), and $2\varepsilon_f' = 0.65$ is the fatigue ductility coefficient.

Maximum shear strain range $\Delta\gamma = (d \cdot \Delta T \cdot \Delta\alpha)/h$

For this package we have:

$\text{DNP} = d = 17$ mm
$\Delta\alpha = 1$ ppm/°C
$h = 0.5$ mm

In the first case, we have a temperature profile from +25°C to +125°C with a cycle time of 40 min. That means a $\Delta T = (125 - 25) = 100°\text{C}$ and a $f = (24 \times 60)/40 = 36$ cycles/day

In the second case, we have a temperature profile is from −20°C to +80°C with a cycle time of 40 min. That means a $\Delta T = (80 + 20) = 100°\text{C}$ and a $f = (24 \times 60)/24 = 60$ cycles/day

Both profiles described above have the same temperature range and same shear strain range. But, they do not have the same mean cyclic temperature T_{m} and frequency.

For the first case:

$T_m = (125 + 25)/2 = 75°C$ and $f = 36$ cycles/day

$c = -0.442 - (6 \times 10^{-4}) \times T_m + (1.74 \times 10^{-2}) \times \ln(1 + f)$

$c = -0.442 - (6 \times 10^{-4}) \times 75 + (1.74 \times 10^{-2}) \times \ln(1 + 36)$

$c = -0.442 - 0.045 + 0.062829971$

$c = -0.424170028$

This means that $(1/c) = -2.3585 \Rightarrow N_f = 9{,}004$

For the second case:

$T_m = 30°C$ and $f = 60$ cycles/day

$c = -0.442 - (6 \times 10^{-4}) \times T_m + (1.74 \times 10^{-2}) \times \ln(1 + f)$

$c = -0.442 - (6 \times 10^{-4}) \times 30 + (1.74 \times 10^{-2}) \times \ln(1 + 60)$

$c = -0.442 - 0.018 + 0.071529205$

$c = -0.388470794$

This means that $(1/c) = -2.5742 \Rightarrow N_f = 22{,}061$

This means that the c value from the first case is smaller (more negative) than for the second case, which means that $1/c$ becomes smaller for the second case. This means that the first case is more damaging in fatigue life.

$\gamma = \Delta U/h = (d \cdot \Delta T \cdot \Delta\alpha)/h = (17\ [\text{mm}] \cdot 1 \times 10^{-6}\ [/°C] \cdot 100[°C])/0.5\ [\text{mm}] = 0.0034 = 3.4\%$

6.6.

$$N_f = C(\Delta\gamma)^{\beta},$$

where N_f is the number of cycles to failure, $\Delta\gamma$ is the plastic shear strain range, C and β are the material constants

Type A: $87 = C \times 0.0866^{\beta}$

$87/C = 0.0866^{\beta}$

$\ln(87/C) = \ln(0.0866)^{\beta}$

$\ln 87 - \ln C = \beta \ln 0.0866$

$$\beta = \frac{(\ln 87 - \ln C)}{\ln 0.0866}. \tag{A.3}$$

Type B: $2{,}250 = C \cdot 0.0101^{\beta}$

$2{,}250/C = 0.0101\beta$

$\ln(2{,}250/C) = \ln(0.0101)\beta$

$\ln 2{,}250 - \ln C = \beta \cdot \ln 0.0101$

$$\beta = \frac{(\ln 2{,}250 - \ln C)}{\ln 0.0101}. \tag{A.4}$$

Put (A.3) = (A.4)

$(\ln 87 - \ln C)/\ln 0.0866 = (\ln 2{,}250 - \ln C)/\ln 0.0101$

$\ln 0.0101\,(\ln 87 - \ln C) = \ln 0.0866\,(\ln 2{,}250 - \ln C)$

$\ln C\ (-\ln 0.0101 + \ln 0.0866) = \ln 0.0866 \ln 2{,}250 - \ln 0.0101 \ln 87$
$\ln C = (\ln 0.0866 \ln 2250 - \ln 0.0101 \ln 87)/(-\ln 0.0101 + \ln 0.0866)$
$\ln C = 0.762489067$
$e^{\ln C} = e^{0.762489067}$
$C = 2.143605164 \approx 2.14$

Put the C value in (A.4) [or (A.3)]:

$\beta = (\ln 2{,}250 - \ln 2.143605164)/\ln 0.0101 = -1.513789687 \approx -1.51$

Answer:
$C = 2.14$
$\beta = -1.51$

Chapter 7

7.1. Due to high aspect ratios (length/diameter) larger than 1,000 they might result in shortenings.
7.2. Electrical shorting test, observation using scanning electron microscopy.
7.3. Coating, heat treatment are the most popular ways to mitigate tin whiskers growth.
7.4. Normally, soldering in nitrogen atmosphere gives better solder joint quality and thereby reliability. So for high reliable applications, nitrogen-assisted soldering is required.
7.5. Lead-free solders have higher surface tension values which makes it more difficult for voids to escape during soldering. Voids can be reduced by increasing the time at which the solder above its melting temperature.
7.6. The solder joints are duller and the wetting is worse resulting in sharper edges and probably some pad area being still visible.
7.7. When combining Bi-containing alloys with Pb, a low melting temperature phase is built, which melts at about 40°C below the same combination without Pb. This phase will result in microstructural instability and reliability concerns.
7.8. Extreme attention has to be paid to the fact that risks are associated with using leaded components mixed with lead-free solder alloys. This will require some new system where companies can distinguish the leaded components from the lead-free ones. Companies have also to be aware on the availability of possible suppliers that can deliver lead-free products, the costs related to the new lead-free products and their lead times.
7.9. Popcorning is the fast expansion of entrapped moisture from the package/component during reflow, as a result of the high temperature that makes the moisture to evaporate and increase the pressure inside the component; this makes the component to bulge and pop and therefore fail.

Chapter 8

8.1. Reliability: the probability that an item operating understated condition will survive for a stated period of time.
Availability: the probability that a system is operating satisfactorily at any point in time, excluding times when the system is under repair.
Derating: a reduction in the ampacity of a conductor due to correction factors. Conductors are rated for a specific set of conditions, and when those conditions change, ampacity must be derated.

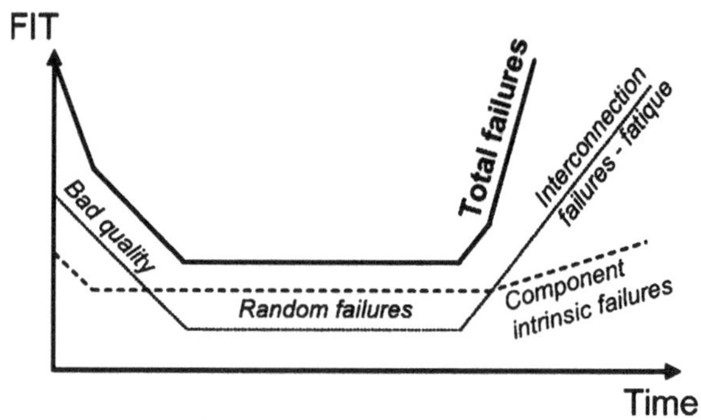

8.2. R (10 years) = 0.96.
8.3. The total hazard rate is a sum of a time-independent constant hazard rate and a time-dependent increasing hazard rate with infant mortality region included. So, the product hazard rate would look more like the so-called bathtub curve shown in the figure. The component intrinsic failures will increase in time, which will increase the hazard rate. If the total product hazard rate is presented, the other failure modes, e.g., the interconnection failures, should be included. This addition increases the hazard rate even further.
8.4. Usually lead-free solders outperform SnPb in thermal cycling tests, the exception being cases where ceramic components are soldered to organic boards (large CTE mismatch) and the thermal cycling is very harsh, such as −40 to +125°C. In real use environment, it is likely that SAC is always more reliable than SnPb.
8.5. Empirical models, physical models, comparing to field data on similar components, comparing results (the other technology's reliability being known beforehand).

Chapter 9

9.1. 0.85738.
9.2. 0.99750.
9.3. 0.90024.

9.4. 0.99988.
9.5. Weibull distribution should be used in the situation that the failure is caused by time-accumulated damage, many competing defect sites but only one causes the failure.
Log-normal distribution should be used in the situation that single weak point would be considered.

Chapter 10

10.1. New technologies are often used in modern technologies and field data and standards relying on best practice for older technologies are often not reliable for predicting and assuring reliability when new technologies are used.
10.2. The first objective is formulated to make sure that a supplier has fully understood the customer's requirements and product needs. The second objective requires that a supplier develops a process that will result in product that meet the customer's requirements and product needs. The third objective addresses the responsibility of the supplier to adequately verify that the customer's requirements and product needs have been met.
10.3. Temperature (steady-state, ranges, gradients, number of cycles), humidity, contamination from production and use, shock and vibration, pressure, radiation, power, current, and voltage.
10.4. Tests designed to assess the risks for early failures are designed to stimulate defects to failures so that defective products can found. The product is then exposed to high levels of stress to increase the probability that a defect is stimulated to cause a failure. This type of test is called environmental stress screening. Highly accelerated stress screening (HASS) is an advanced form of ESS where data from a previously performed HALT test is used to select stress levels as high as possible. The main purpose is to determine the percentage of the products that are defective and what type of defects do they have.
10.5. Tests for assessing the risk for failures due to aging and wear out must be designed to simulate the true failure mechanism in field conditions, which is they must be based on physics-of-failure. Identification of the crucial failure mechanisms requires knowledge of the life cycle loading conditions, product architecture, and manufacturing processes. Normally, this type of test must be accelerated to achieve test results in a reasonable time. This can be achieved by increasing the frequency of the occurrence that causes failure or by increasing the severity of the conditions causing the failure. If the acceleration factor is known for the test, the failure rate in field conditions can be assessed. Normally, one accelerated test needs to be designed for every specific failure mechanism.

Chapter 11

11.1. Higher shear rate leads easier to brittle failure, while lower shear rate leads to ductile rate.

11.2. Thermal shock generates extremely high temperature changing rate. It normally leads to brittle failure while thermal cycling can induce fatigue and creep-related failure with smaller temperature changing rate.

11.3. Displacement $= \dfrac{1}{2,400\ \text{line/mm}} \times 13\ \text{line} = 0.005417\ \text{mm}$.

11.4. We can use a differential scanning calorimeter (DSC), which is a widely used thermal characterization technique. DSC measures the temperatures and heat flow associated with transitions in materials as a function of time and temperature. The technique provides qualitative and quantitative information about physical and chemical changes that involve endothermic or exothermic processes or changes in heat capacity.

The glass transition event occurs when a hard, solid, amorphous material or component undergoes its transformation to a soft, rubbery, and liquid phase. The temperature at which the glass transition occurs is known as the glass transition temperature (Tg).

DSC is the most common approach used to measure Tg, which is observed as an endothermic stepwise change in the DSC heat flow or heat capacity. The magnitude of the step change at Tg is very dependent upon the sample's chemistry as well as physical conditions, such as crystalline content and orientation. At high levels of crystallinity, the glass transition may not even be detectable by DSC.

11.5. Accelerated aging is a testing method used to estimate the useful lifespan of a product when actual lifespan data is unavailable. This occurs with products that have not existed long enough to have gone through their useful lifespan. The test is carried out by subjecting the product to unusually high levels of stress, designed to mimic the effects of normal use. Mechanical parts are run at very high speed, far in excess of what they would receive in normal usage. Also, the device or material under test can be exposed to rapid (but controlled) changes in temperature, humidity, pressure, strain, etc. Humidity and temperature testing, thermal shock and thermal cycling testing are examples of the accelerated aging tests.

Index

FSC
www.fsc.org
MIX
Papier | Fördert gute Waldnutzung
FSC® C083411

Kolibri 360 GmbH
Ferdinand-Jühlke-Straße 7
99095 Erfurt, Deutschland
produktsicherheit@kolibri360.de

The New Synthese Historical Library

Texts and Studies in the History of Philosophy

Volume 76

More information about this series at http://www.springer.com/series/6608